Michael Kuhn

How the Social Sciences Think about the World's Social

Outline of a Critique

BEYOND THE SOCIAL SCIENCES

Edited by Michael Kuhn, Hebe Vessuri, Shujiro Yazawa

ISSN 2364-8775

1   *Michael Kuhn,* Shujiro Yazawa *(eds.)*
Theories about and Strategies against Hegemonic Social Sciences
ISBN 978-3-8382-0586-1

2   *Michael Kuhn*
How the Social Sciences Think about the World's Social
Outline of a Critique
ISBN 978-3-8382-0892-3

3   *Michael Kuhn, Hebe Vessuri (eds.)*
The Global Social Sciences
—Under and Beyond European Universalism
ISBN 978-3-8382-0893-0

4   *Michael Kuhn, Hebe Vessuri (eds.)*
Contributions to Alternative Concepts of Knowledge
ISBN 978-3-8382-0894-7

5   *Kazumi Okamoto*
Academic Culture: An Analytical Framework
for Understanding Academic Work
A Case Study about the Social Science Academe in Japan
ISBN 978-3-8382-0937-1

Michael Kuhn

HOW THE SOCIAL SCIENCES THINK ABOUT THE WORLD'S SOCIAL

Outline of a Critique

ibidem-Verlag
Stuttgart

Bibliografische Information der Deutschen Nationalbibliothek
Die Deutsche Nationalbibliothek verzeichnet diese Publikation in der Deutschen Nationalbibliografie; detaillierte bibliografische Daten sind im Internet über http://dnb.d-nb.de abrufbar.

Bibliographic information published by the Deutsche Nationalbibliothek
Die Deutsche Nationalbibliothek lists this publication in the Deutsche Nationalbibliografie; detailed bibliographic data are available in the Internet at http://dnb.d-nb.de.

∞

Gedruckt auf alterungsbeständigem, säurefreien Papier
Printed on acid-free paper

ISSN 2364-8775

ISBN-13: 978-3-8382-0892-3

© *ibidem*-Verlag
Stuttgart 2016

Alle Rechte vorbehalten

Das Werk einschließlich aller seiner Teile ist urheberrechtlich geschützt. Jede Verwertung außerhalb der engen Grenzen des Urheberrechtsgesetzes ist ohne Zustimmung des Verlages unzulässig und strafbar. Dies gilt insbesondere für Vervielfältigungen, Übersetzungen, Mikroverfilmungen und elektronische Speicherformen sowie die Einspeicherung und Verarbeitung in elektronischen Systemen.

All rights reserved. No part of this publication may be reproduced, stored in or introduced into a retrieval system, or transmitted, in any form, or by any means (electronic, mechanical, photocopying, recording or otherwise) without the prior written permission of the publisher. Any person who does any unauthorized act in relation to this publication may be liable to criminal prosecution and civil claims for damages.

Printed in the EU

Table of Contents

Acknowledgements 7

Preface 9

Why a theory about social sciences? 11

Chapter A: The world's social in social science thinking 23

    Social sciences detect the world's social beyond the national biotopes 23

    ...by assembling theories about nation state social biotopes 29

    ...off-thinking the world's social 33

    ... reflected on through the nation state constructs 45

    ...ever critically measured against idealized nation state rationales 49

    The universalization of social science thinking 53

    ....completing the globalization of social science theorizing
as a multiplicity of scientific patriotisms 57

    From Marx to Heidegger: Critical theorizing in the anti-colonial
movements—self-purified for constructive imperial nation state views 64

    ...opposing a monopoly on spatiological thought in the "centres" 70

    .... liberating global social thought
from scientificy for creating patriotic theories 76

    ... and anti-scientificy to practice global social sciences 82

    From patriotic to imperial social science thinking 84

    Nationalism: A service for imperial social science theorizing 91

    ...thought back by alternative imperial social science models 96

    ...critiquing an unequal knowledge imperialism 99

    The world's nation states serving the world's mankind 103

Chapter B: Categorical essentials of disciplinary thinking ... 105

    The cognitive architecture of disciplinary thinking 107

    Essential concepts founding theorizing
in the classical social science disciplines 115

        Anthropology—Regimen as the demand of man's nature 116

        From anthropological thinking to cultural theories—nation states
as cultural artefacts completing man's unfinished nature 126

        Economic thinking in the social sciences—The bane of scarcity 131

        Sociological thinking—The blessing of the "community" 136

        Political theory—political power for the politically disempowered 144

        Psychological thinking—the mythologization of the mind 148

Essentials of social sciences disciplinary thinking 157
    1.   The common cognitive lie founding
        the categories of disciplinary thinking 157
    2.   The shared metaphysical nature of the disciplines
        and their speculative way of theorizing 160
    3.   Disciplinary social thought cannot think other
        about the social but as an idealized nation state social 163
    4.   The categorical essentials: Critically
        affirmative and idealistically domesticative 164
    5.   The world's social in disciplinary thinking—absent 166

Chapter C: The social science approach to scientific thinking—advancements of teleological theorizing 169

The social science mode of thinking—
cognitive operations of a methodological idealism 173

Social sciences theorizing about social science thinking 179

Why teleological thinking must be the nature of thinking 187

The stigma of the natural sciences—
and the self-destruction of an envied hero 191

The envied hero 192

...and his self-destruction 196

The decline of scientific knowledge towards ephemeral knowledge 200

Chapter D: The discourse about and the progress of social science knowledge 209

The discursive creation of acknowledged true knowledge 210

Paradoxes of acknowledged knowledge
in the global social science discourse 218

Global discourse about acknowledged knowledge
ruling social science theorizing 220

Arguing about the position national knowledge bodies hold 224

The progress of acknowledged knowledge 229

How to create a globally shared truth ruling global theorizing 235

The ephemeral progress of ephemeral knowledge 255

Chapter E: Going beyond the social sciences 263

Postscript 269

Acknowledgements

This book is an outcome of the project "Social Sciences in the Era of Globalisation", funded by the Calouste Gulbenkian Foundation in Lisbon.

It was only possible to write it thanks to all the controversial discourses with numerous colleagues around the world committed to discuss the issue of the above project.

Preface

Outlining a theory about the social sciences and critiquing social science thinking as thinking that creates false theories must fail—according to social science thinking. Thanks to their concept of critique social science thinking is immune against a critique that critiques false thought. Since social sciences deny that theorizing about the social—and since T. Kuhn's book interpretation through the social sciences also theorizing in natural sciences about the nature—are able to create right theories, there cannot be any critique that a theory is a false theory. Social science theories may well be critiqued, but this critique is not a critique of any false theory, but a critique that argues against all the ex-ante definitions any social science theorizing requires to reveal all its ethical, ontological and methodological choices theorizing must make, choices of what a theory is about and how a theory intends to think about any object of thinking. It is not at all the case that social sciences are not inviting critique. However it is a critique that does not allow critiquing theories as false theories. It is the ex-ante definitions and choices which open a wide field for critique about all those choices and definitions, however, a critique that a theory creates false thought is no option in this kind of critique, since any theory can be only false in relation to the choices and definitions it must make for theorizing. Nevertheless, since even all these tautological cognitive operations of a critique of in this sense false thought in social sciences, like any scientific thought, require to be made plausible, because they are cognitive operations of scientific thinking, they need to be founded with reasons. Though some social sciences meanwhile simply present data and consider this as having created a founded theory, creating scientific thought cannot just state what a scientist thinks without any arguments reasoning why he thinks what he thinks. And this is the weak point in this very immune system of social sciences critique against a critique of false thought, because even the immune system of criticism, immune against a critique of false thought, must argue why there is no false and right thought and why the thought that there is no right thought making social science immune against the critique of false thought, is right thought. This is why it is worth trying to critique social sciences thought where they create false theories, though such things like false thought—according to the social sciences theories about social sciences theorizing—do not exist in social science theorizing.

Why a theory about social sciences?

'Globalization' is—according to the social sciences—not only *the* essential, crafting the contemporary social life, but also the reason urging the 'internationalization' of the social sciences. This statement contains at least two false thought and one odd confession, bought with a discreet myth, a self-deception about the social sciences.

To start with the confession: The fact that contemporary social sciences are quite heavily engaged in discussing the need to internationalize thinking about the social, is as odd as telling, since it confesses that thinking about the social beyond individual nation state socials at least until now was not at all a topic for social sciences.

This confession contains the false thought, that the today's discovered "globalization" of the social means that there was no global social before social sciences found out that there is global social. It is only the discovery for the social sciences that there is a social world beyond what their theories are normally about, which have obviously so far been focusing on any nationally confined objects of thinking, just as if there was no global social in the previous—colonized—world's social. It needed the global spread of nation state socials to make social sciences detect a 'globalization' of the social beyond the individual national social entities. A social world that is not a world of nation state socials for social sciences obviously is no global social. Only the postcolonial transformation of the world into a world of nation states makes the social sciences realize that there is a world social beyond the secluded national socials.

And this, recognizing a global social after the world became a world of nation states is a discreet myth of the social sciences about the social sciences, since the social sciences did very well know a social world beyond the nation state socials; they even created a particular discipline, anthropology, that was and is in charge of the "non-civilized" social, that is all socials which were no nation state social.[1] It is though telling that with the exception of anthropology, reserved for the non-nation state socials, a world social did not exist for the social sciences until the world was made a world of nation state socials.

The notion 'globalization' articulates this image of a detected world's social, social sciences identify with the agglomeration of individual nation state socials, just as if a world of nation state socials was the final completion of the world's social nature: 'Globalization' is the world-wide spatial spread of something, which does neither have any subject that makes anything global, nor any object, any what that is globalized, nor does the notion reveal any forces or reasons, which are responsible for this mysterious global spread of a subject—and objectless something, just as if the world finally became what it ever quasi naturally was.

From the fact that social sciences are advocating the need for internationalizing social science thinking today, one must in fact conclude, that it took 200 years to make the social sciences in the imperial world notice that there was and is a world beyond the nation state social of the imperial nation states. As if the establishment of the imperial countries was not the result of the colonial subjugation of the world by the imperial nation states, exploiting the colonized world and setting the economic basis for the economic wealth and the global power of the imperial world, as if the world of nation states and their imperialism subordinating the world under their command was not the way the global social was and is *made*, the social sciences, more precisely the social sciences in the imperial nation states, apparently only noticed that there is a social world outside of their national social biotopes, after their national science policies detected science as a means for the global competition about economic growth and political power and therefore forced the sciences to pay more attention to the world beyond the national socials. The fact that it was indeed national science policies that "encourage" social sciences to work internationally is telling. It obviously needed and needs such political "incentives" to make social science detect an era of "globalization", just as if the

---

1 Castro, N., (2013) Isn't Anthropology Already a Multiversalist Discipline?, In M. Kuhn and K. Okamoto (eds.) *Spatial Social Thought, Local Knowledge in Global Science Encounters*, Stuttgart, ibidem, p 33

world before consisted only of their secluded national social biotopes.²

For social sciences, their discovery of a world's social, therefore, still is rather the detection of an exotic elsewhere. Despite all the debates on the need to internationalize social thought, the main social sciences theory production is yet to be bothered by such debates and continues their routine work creating knowledge not only confined to nation state socials, but knowledge constructed through the perspectives of the peculiarities of individual nation states socials, namely those of the imperial world. Still, thinking beyond a nation state social, is a rather exceptional and adventurous scientific undertaking, despite all the debates about internationalizing social sciences.

Not that much inspired by their own intellectual curiosity about what is happening in the world, not to mention any theoretical needs to understand any social phenomena in an imperial world, social sciences, asked and pushed by the political elites, of course, not to pay more attention to the world beyond their national social islands, but to take part in profiling the national knowledge resources as an appealing resource for the global capital, calling this the need of "globalization" for globalizing social sciences, reveals that not only the existence of a *world* of nationally constructed socials was a new phenomenon for social sciences, namely in the imperial world.

Consequently, the international or global knowledge responding to their new discovery of a world's social still continues to consist of nationally constructed knowledge: The main way that comes to the mind of social science thinkers to look at the world's social, is to compare their knowledges about the confined national socials. It seems, social sciences, confronted with their enforced detection of a world's social beyond their theoretical constructs of secluded national social biotopes, apparently do not know anything else but

---

² Needless to say, that the social sciences in the imperial supervisor of all imperial countries, the US-social sciences, the detection of a social world started much before the today's debate about "internationalising" social sciences with the establishment of "area studies". The dreamy illusion of a secluded nation state social is a privilege mainly of social sciences in those imperial countries, in which only the detection of science as a means of global economic competition caused the wake up call to detect the world's social as an international social.

theorizing about the world's social other but as a multiplicity of national socials and cannot interpret thinking about the global social other but accumulating nationally constructed social thought. Just as if they would simply not know how else they could theorize about a world's social other than assembling nationally confined social thought.

However, there are a few social sciences, mainly from the "developing" world, which insist that there is a view of the world's social beyond the illusionary construct of national socials and which very well know that constructing a world of secluded national socials, is an imaginary image of the social sciences in the imperial world.

Such an odd imaginary construct, thinking about any social only as nationally confined biotopes, could certainly hardly happen to scientists in those parts of the world, where the dependence of any aspect of the very national social reality from imperial countries would hardly allow such a "zombie"[3] science, presupposing this national social as a secluded national biotope unaffected by the world's social, and that detects the world's social only once it became a world of nation state socials.

From their point of view, creating such an illusionary knowledge view on the social is too odd when thinking about national socials, which are—though in a rather formal sense—also nationally constructed, but are national socials where the political and economic substance of these nation state socials are entirely under the command of and for the service of the imperial nation state and do not allow the illusion of individual nation states as the exclusive agent crafting a secluded social, as the social sciences in the imperial world want to believe.

However, rather than being irritated about the explanatory abilities of such illusionary social thought, it seems that theorizing in any social sciences anywhere simply does not know what knowledge that is not constructed about nation state socials and not seen through the parochial view of nationally confined theorizing, could be at all about. It seems that it is the nature of social sci-

---

[3] With this observation of a "zombie science", Beck is actually very correct. His further elaborations on this observation end up replacing thinking about national biotopes with the alternative of thinking about the world's social as genuine imperial thinking. (See section I) http://www.ulrichbeck.net-build.net/index.php?page=cosmopolitan

ence thinking that thinking about the social must be nothing but thinking about and through the constructs of nation state socials and that the only way for social thought in the social sciences to recognize the world beyond national socials is, therefore, the aggregation of nationally constructed social thought.

Hence, despite of the difficulties, to think the former colonial nation state socials as secluded national biotopes, applying social science thinking to the former colonies the social sciences in the new nation states also think about the world's social through such national constructs. Indeed, observing the global debates about the globalizing social sciences, their main arguments about "scientific power", the "in-equalities", "scientific imperialism" and alike, are also always discussed along nationally constructed entities, may this be a "North" versus a "South", local versus global, Eurocentrism or Occidentalism, rather than having any hesitations about the preoccupations of ever nationally constructed global social thought, always assuming the national social could be understood as nationally constructed social. Practicing the newly detected mission to "globalize" social thought, under the regime of social sciences is ever interpreted as the need for more "local" thought, more nationally constructed theories, to take part in the creation and debates about global social thought as an "equal" contribution to the assemblage of nationally constructed theories.

Thus, strikingly, the more social sciences strive for internationalizing social thought, the more they devote social thought to the world's social, the more they stress the need for thinking about national socials, not only as their unit of analysis, but as their way of thinking about the world's social as an collection of theories about national socials. To create global social thought, social sciences not only think about their nation state socials, they understand the creation of global social thought as to look at their national socials through an exclusive national perspective that only works for thinking about the social within these nation state social islands and, thus, making it even impossible, to share and assemble all those parochially constructed knowledges across these clandestinely constructed local/national knowledge bodies.

Rather than questioning the national knowledge constructs, globalized social science thinking that confines thinking to individual nation state socials, in a world consisting not only of a multiplicity but also of an essential diversity of nation state socials, introduces and insists on a distinction between the epistemological impacts of the many 'wheres' of knowledge. To join global social thought under the regime of the social sciences, social sciences in the decolonized world create all kinds of spatially distinguished knowledges, local, global, southern, northern, universal and alike knowledges—and wonder about a reciprocal ignorance about what is going on beyond their individual secluded "wheres".

Not only is the contemporary detection of a global social and the illusionary way of theorizing in the social sciences about the world's social, a way of thinking that seemingly is not able to think about the world's social other than through constructing a world of secluded national islands, even when the social reality in the ever "developing" world obviously disobeys this way of thinking about the social are enough reason to urge thinking about social thought under the global regime of the social sciences and to find out what the nature of social science thinking is

Yet, there is another observation regarding the theoretical substance of the knowledge social sciences create since now more than 200 years of social science theorizing that also urges to think about social thought under the global regime of the social sciences and this is to wonder what the influence is that the always critical knowledge social sciences create has on the social world?

What do the social sciences let us know about the world's social, a world's social which is, since the social science create knowledge, a world of war and the coexistence of growing wealth and growing poverty[4] and it is this for more than 200 years of social science thinking? Certainly, one cannot make the social sciences responsible for what is happening in the world: the globe, social sciences call "modernity", a place characterized by war, poverty and wealth. Is there any place on the globe that is not involved in any wars? Is there any place in the world, where the growth of wealth does not

---

[4] It does not need 200 years of social science research to understand that at least the opposition between growing wealth and growing poverty, is the practiced contradiction of the capitalist production of wealth, for which any consumption is not only not the aim of the production of wealth, but in which consumption is questioning a type of wealth that measures its success in abstract values and that therefore treats the consumption of utility values as a conditioned means and deduction from their success criterion. Only a view into the current newspapers discussing how to make a country like Greece more productive is full of examples for how the production of capital in the first place urges and needs poverty.

co-exist with to the growth of poverty? Certainly, war, wealth and poverty are the essentials of "modernity" and they have been this for more than 200 years.

And, not to forget, for more than 200 years the social sciences think about the social world with an army of professional thinkers and create ever critical theories. Has the knowledge they have created and create at least helped to make anything to the better or at least to reduce wars and the co-existence of wealth and poverty? Obviously, not, rather the opposite is the case: There are increasingly more wars and there is a growing gap between wealth and growing poverty and this, where ever, across the whole world.

Again, one cannot blame the social sciences for this, after all, knowledge is knowledge, but what is the impact on the social world of all the mainly critical knowledge these armies of professional thinkers create about the social world? Nothing much, one must conclude, if one assumes that social sciences aim at reducing wars and poverty[5]. And since it is also sure, that social sciences do not propagate war and poverty, but rather critique them, one must raise the question, what the impact of social thought under the regime of the social science, what the impact of all the critical knowledge the armies of social scientists create about the world's social since 200 years, is after all? Still, it is the social science theories not only providing their societies with the knowledge they have about themselves, it is also this knowledge the society acquires through education and it is this education system from which they recruit all the governing positions. What is the role social sciences play in the reproduction of the nation state societies and their market economy, why does 200 years of researching the world of nation states and market economies and all the critical theories about them obviously have no impact on a world ruled by wars, wealth and poverty—again, assuming that social sciences not only critically argue about, but really aim with their knowledge at reducing wars and poverty, as in our example from Skinner, not to mention

---

[5] As an example Skinner: "In trying to solve the terrifying problems that face us in the world today, we naturally turn to the things we do best. ...Threatened by a nuclear holocaust, we build bigger deterrent forces and anti-ballistic missile systems. We try to stave off world famine with new food and better ways of growing them.... We can point to remarkable achievements in all these fields...But things grow steadily worse,..." Skinner, B.F., (2002) *Beyond Freedom and Dignity*, Indianapolis, Reprint Skinner Foundation, p 1

at abolishing both. Or is this anyway already a wrong assumption, considering how the world is developing—despite the critical social science knowledge? Or is it because of all its critical knowledge or neither nor?

Distinguishing in the following reflections in this book about social thought under the global regime of the social sciences between social thought and the social sciences, theorizing about the social, implies, in fact, that the social sciences are only a particular historical form of social thought. Indeed, the reflections in this theory about the social sciences, about global social thought[6] under the regime of the social sciences, hold that the way social sciences think about the world's social not only results in particular theories about the world's social, but that the way they reflect on social phenomena is a very particular way of theorizing, typical for how only social sciences theorize and typical for the role the knowledge they create plays in the social world. In fact, this implies that social sciences are only a particular interpretation of theorizing about the social, not at all congruent with the nature of scientific thinking, and that it is only the social sciences way of theorizing that is responsible for the phenomena only social science thinking creates and that is responsible for the knowledge the social science approach to social thought contributes to the world's social, a world ruled by wars and poverty.

Interpreting the historical form of any social human practices as coinciding with their nature might be understandable from the practical point of view of a practitioner, who is too much caught by the practical necessities of what he is doing, to reflect upon such ontological issues. However, if scientific thinking about the social is identified with the way of thinking in the social sciences, it indicates an irritating ignorance of the very social sciences about their particular format of thinking. It does this the more, if one not only remembers that thinking about the social had historical predecessors theorizing about the social, among which a number of essen-

---

[6] Using the term "global" social thought only names the object of social thought, it is just another phrase for theorizing about the world's social, a social crafted by the world's whole nation state social. It does not indicate, as it does in the contemporary social sciences discourses, a way of spatial theorising, a concept that is discussed later in chapter 3 in this book.

tials, characterizing the particular nature of social sciences, were unknown, such as the plurality of social sciences.[7]

In the first place and on the first glance one could indeed notice that the historical predecessors of scientific thinking about the social, thinking divided in scientific disciplines did not exist and only occurred with the emergence of the social sciences.

One could, secondly, also easily notice that thinking about any social phenomenon was thinking about this phenomenon and that this thinking was not confined to any spatially constructed unit of analysis, mostly nation states[8], as this is the case in the way, social

---

[7] An example of the ignorance among the social sciences about the very nature of the social sciences is Wallerstein's famous book, known under the title "Unthinking Social Science". Wallerstein, accurately describing the historical emergence and developments of the social sciences, the emergence of the individual disciplines, and, apparently assuming that his mere descriptions of what happened was anything else but merely observations of a typical way of sociological thinking, the practitioners of thinking about the "real facts" (Comte), obviously consider this description as an explanation for what happened and why it happened. It is striking, how a scientific thinker, thinking about the emergence of the social science disciplines, the battles between them about their subjects matters, tries to avoid even raising any questions about issues Wallerstein does mention and which so much call for explanations; for example the question, why social thought under the social sciences divides theorizing into a multiplicity of different disciplines. There is not only no question nor any answer for this, and one must conclude that Wallerstein considers his descriptions, lacking any theoretical concepts, as the same as explaining *why* all this happened. Social thought for social scientists must be so much the same as the way social sciences practice social thought that the need to explain social sciences and their peculiarities does apparently not occur when social sciences theorize about the social sciences, and this, remarkably, *while* theorizing about them. Wallerstein, I., (2001) *The Limits on Nineteenth-Century Paradigm, Unthinking Social Science*, Second Edition, Philadelphia, Temple University Press.

[8] The nation state is the enforced community, a set of enforced constructs of humans defining them and their live with a monopole of political power under the nation states rationale and objectives, forcing with this power monopole to perform live as conflicting interests and to carry out these conflicting live rationales of humans under the rules of citizenship, to use the conflicting interests of the nation state creatures, citizens, as their means to strive for their live aims.
As it will be shown in this book, it is the elementary mistake of social science thinking to separate the nation state and all its social constructs from an imaginary national entity, in order to create the many wishful ideas what this imaginary entity may be, and to measure the real nation state—just as if the nation state was anything else but the definitions of what nation state constructs are, defined with the nation state's political power—against such wish-

sciences think about any social phenomenon. None of the classical theoreticians such as Kant, Hegel, Marx, Smith or Hobbes constructed theories about an issue spatially confined to a particular country, such as confining a critique of rationalism to a critique about theorizing about rationalism in Germany, to mention only the example of Kant's work. And, needless to say, such theories contained reflections about modifications to the topic they reflected on, may they be historical, local or any conceptual diversities of the issue they discussed—just as Marx and Smith did it while working on theories about capitalism, distinguishing phenomena of capitalism in England, Germany and in India, to only mention the example of variations—not of theories about capitalism, but of capitalism.

Apart from such obvious historical differences between social thought in the classic philosophies and the social sciences, thinking about the social and, as this book does, discussing how the social sciences last but not least currently reflect on the—global—social, face a number of paradoxes, which could at least prompt the question of why theorizing in the social science creates such odd phenomena, odd phenomena that should raise the attention of social thought and motivate them to reflect on how social thought under the—global—regime of the social sciences works.

This book discusses, why all these oddities of social science theorizing encountering a world's social and why the dubious impact all the critical social science knowledge has are not a 200 years lasting accident, but the inevitable result of the particular way social sciences theorize about the social, a necessity of the particular nature of how social sciences think about the world.

---

ful ideas this imaginary entity is aiming at. It is this false separation into two entities and, built on this separation, a discreet comparison of both, that founds social science thinking and its critical nature and as its nature of critique. What this other imagined nation entity may be, any national entity beyond the real nation state, that is only the enforced construct the nation state is, is what constitutes social sciences thinking across all their disciplines. Proving this is the topic of the following reflections about the social sciences.

It discusses this in five chapters:

A. The world's social in social science thinking
B. Categorical essentials of disciplinary thinking
C. The social science approach to scientific thinking—advancements of teleological theorizing
D. The discourse about and the progress of social science knowledge
E. Going beyond the social sciences

Chapter A
The world's social in social science thinking

Social sciences detect the world's social beyond the national biotopes...

Social sciences are seriously challenged if they are requested to think beyond their national biotopes, especially in those countries from where they originate. One hundred and fifty years of colonizing the world, exploiting the world to build the basis for their economic and political global reign over the world, and another half century after the post-colonial US model of imperialism, transforming the former colonial part of the world into players on the global battlefield, making the whole world into a world of nation states, all substantially constructed along the US nation state rationale, a world of nation states divided in—competing—imperial powers and rather more formal nation states, all bound under the supervision of the US empire to serve and to make their global power dependent on the benefit they gain from the growth of global capital they serve, it took social sciences thinking in the imperial countries another 50 years to realize that there is a world outside of their nation states, a world they feel they should no longer ignore. In particular, the social sciences in the imperial nation states call for an internationalization of social sciences—and to inter—nationalize (sic) theorizing about the world's social, performed by comparing nationally constructed social thought.

Strictly speaking, as said before, one should say that it was not the social sciences that detected that there was a world beyond the uniqueness of their nationally constructed societies. It was the national science policies in the imperial nation states, later followed by the dependent world that forced the social sciences to shift their thinking from their nationally confined unit of analysis towards other nationally constructed societies, at least to those, science policies detected and detect any political or economic interests in. In fact, the selection of the nation state socials attracting more attention by social sciences are those, in which the imperial world has any economic or political interest, may this be because they are under the exclusive grip of another competing imperial power. The newly detected interest of the Europeans in Latin America questioning the scientific monopoly of the US, or the interest of Japan in South East Asia, promoted by accordingly directive funding programs, may serve as examples.

And even this is not the full truth. Really strictly speaking, it was not even the science policies in the imperial world that detected the world beyond their territories as a topic for science. It was the globally acting capital which throughout its history found and finds the confined territories of their nation states as confining their business and pulled down any local barriers making the world into a means for the growth of capital. And, since the global capital discovered with the emergence of the new technologies science as a whole as a crucial means for their competitiveness, science as a whole, including the social sciences, raised the interests of the economy and, as a consequence, the interests of the nation state, getting science under their political control for these newly defined objectives of science policies. Science policies detected the new interests of capital in science and served these new needs, awakened science, at least the nation state driven parts, from its ivory tower and transformed nationally directed sciences into a global knowledge market, a global economic resource, once global capital detected science as a major means for their global business. It is since then that the global capital uses the world's knowledge as an economic resource, reorganized by policy reforms to serve the rationales of the global political and economic players.

To do this, national science policies transformed their sciences towards one of the major politically supervised economic resources of nation states, offered to the global capital and forced their national sciences to compete on an international knowledge market about the attractiveness of the national science markets, sociologists emphatically like to call "national science communities". Not only have the institutional settings of sciences therefore been adjusted to the needs of the world's business demands, forcing knowledge to obey the rules of a global commodity; the whole set of categories, in which in particular social sciences think about the social have accommodated themselves to think in categories, which reflect the transformation of the whole former world of science and education, until then insisting on their independence from politics and business, into an subsection of the national economic infrastructure.⁹

What all the newly emerging debates promoting the globalization, the internationalization or the cosmopolitanisation of social sciences do not want to know, is the reason why social sciences should shift social thought towards a global social, all implying the assumption that they are not global. Arguing that it is the current globalization of the world's social, that requires the globalization of the social sciences, presents the false ideas of the social sciences, that they so far were not global and became global due the new global nature of the global reality, which is just as false.

They are false, firstly, because they argue that the world's social reality was not a global social in the colonial world period and secondly, that the social sciences so far have not been global. And these two false statements about globalizing social sciences are already telling about the nature of social sciences: Firstly, monopolizing social sciences in the imperial world and excluding the colonized world, the world without nation states, from social science thinking in the classical social science disciplines *was* the very way of global social science theorizing. And, secondly, because the colonized socials were no nation state social, their social was, indeed, no topic for those social sciences, which reflect about the national biotopes, the classical social science disciplines, and which were reserved for thinking about nation state socials. Therefore, within the very nomenclature of the social sciences, the colonized world, the world without nation states, was a case for Anthropology and thus, in this very way, they *were* a very topic of the social sciences.¹⁰

---

9 In Europe and from there also spread beyond Europe, this adjustment of science to a service for global business was set into force the "Bologna process". Any single category used in this "process", processes the idea making the sphere of knowledge a means for global competition about knowledge as a means for economic growth. As usual, the social science world sensed a "change" and eagerly searched for accommodating theorizing to this change. The accordingly constructed theory providing the buzz words for this new policy coined this as a "knowledge based economy", which further on guided the adjustment of all social science theories towards the categories this economic theory created for future social science theorising. (See section D)

10 R. Connell rightly mentions that during the historical battles among the disciplines about the topics of disciplinary thinking, also the classical disciplines reflected on the non-nation state socials in the colonized world until they found the nation state social as their native topic and devoted thinking about the non-nation state natives to the a disciplines constructed for thinking

The fact, that the de-colonized countries, once they gained the status of a nation state, concluded from the monopoly the imperial world held on classical social sciences disciplines, that it was an opposition to the ways the social sciences reflected on the de-colonized social to implement social sciences in their countries, was and is one of the tragic errors of an opposition, that wants to be part of what it opposes. A view on the unbroken reign of the racist theories from the imperial world about the new decolonized nation state socials, could signal this wrong conclusion as this tragic error.

The fact that the social sciences reflect in nationally constructed entities across the globe about the national socials, reserving a particular disciplines for the non-nation state social in the colonies, is and was their very way of a very inter-national reflexivity. And, indeed, the current practices which have shifted social science theorizing towards the rightly called—inter-*nationalization* of social sciences—, not abolishing or at least questioning the national outlook, but extending these very national views on the world's social towards other nationally constructed socials, confirm that the least they were and are interested in, is to think about the world's social, if at all, other than through an assemblage of nationally constructed theories. Social science thought continues thinking in secluded nation state social units and if they are requested to think beyond these social biotopes, they compile and compare these biotope-like, nationally constructed theories.

As if the world constructed from nation states was not a way to construct the *world*, a world's social consisting of a multiplicity of nation state socials and of the rationales of nation states, which all consider the territories, the people and the natural resources in these territories as means to combat other nation states of the very same kind over using each other across the world for their economic growth and their political power over each other, all nation states, striving to subordinate others of the same kind across the world under their political and economic command, social science thinking considers the individual national socials as secluded biotopes, exclude the "outside" that mainly crafts the "inside" across the

---

about the non-nation state socials: Anthropology. Conell, R. (2007) *Southern Theories, The global dynamics of knowledge in social science*, Cambridge, Polity Press

world, just as if reflecting about any nation state social would allow one to understand this secluded social, not to mention, that the agglomeration of nationally constructed social thoughts was the same as theorizing about the global social.

Social science thinking not coincidentally once named "*Staatswissenschaft*", presupposes an image of the world of nation states, in which their humans inhabit secluded islands that are not affected by what is going on beyond them. Thinking about the "beyond" is no topic for social sciences; they are the subject of a sub-department of political science, reflecting—if at all—on foreign affair policies and of Anthropology, today more and more replaced by "intercultural studies", acknowledging after more than 50 years of a de-colonized world other nation state socials as socials generalizing racist reflections about "others", so far reserved for the non-nation state socials, now to the whole world's nation state socials.

Social sciences seemingly derive from the fact that caring and thinking about other nation states is the business of a selected and limited number of humans, the political and economic elites, that humans' life *within* these biotopes is not mainly made by international relations of nation states, that is their battles about political and economic power, an illusion created by the sovereignty of nation states over their people and territories, an illusion that can hardly occur in nation states where this sovereignty is only a formal sovereignty. Inhabitants of nation states in those parts of the world, in developing countries, that served and serve through the exploitation of the products of their work and with their natural resources for the growth of wealth in the imperial countries, do not only know that nation states are no secluded islands and can easily experience that their life is mainly defined by those who exploit their work and their resources. And they do not only not share the illusion that the sovereignty of nation states over their people makes people's life unaffected from other nation states. They also do not see the need for globalizing social thought, as the social sciences in the imperial countries, the beneficiaries of the world of nation states do since they detected that there is a world of nation states beyond their own nation state in which they detect their political and economic interest, they since then call an era of "globalization", just as if the whole world was a world only most recently ruled by the imperial nation states.

As if the history of nation states, more precisely the foundation of the imperial nation states, namely those in Europe, and their economic wealth, their genuine economic accumulation of capital, was not the result of expropriating the former colonized world, a wealth they use until today to dictate the terms of business and power in a post-colonial world, social science thinking discovers with the help of a hint from their political and economic elites, that there is a world beyond the biotopes of the imperial nation states, finding a world of nation states that was completed by the former colonized part of the world.

However, from the point of view of social sciences and their routine work, especially those in the imperial world, there was and, looking at how they detect the world beyond the imperial countries social, there is no need to pay much attention to the world other than theorizing about the individual nation state socials. International social science is still an exceptional adventure and the majority of the social science armies across the world's nation states still confine theorizing to the secluded nation socials, mainly those in the imperial world.

Just like the inhabitants of the national social entities do not need to know any much about the world beyond their national social to get on with their life as nation state citizens, with the exception of a few specialists dealing with the other biotopes, a few business people and politicians, the professional thinkers of these societies are not seriously interested or engaged in thinking about the social beyond their national social islands—not to mention if and how the global interaction of nation states craft the social life within them. Next to the debate about the need to internationalize social science theorizing the majority of social sciences can carry on with the illusion on which their theorizing is constructed, that is that any individually national social is what social science theorize about convinced to thus understand the nationally confined social.

Ignorance, exoticism and demonization are not bad attitudes of social scientists, namely "Western" social scientists, but apparently an epistemological presupposition of social science thinking, which considers the secluded nation state social as their topic of reflections and, if at all, the outside world as the complementary topic social sciences in other national biotopes they need to care about, to arrive at inter-national social sciences as the assemblage of nationally constructed knowledge bodies.

As a result, after 200 years of social science theorizing about the world's social and the more recent shift towards globalizing social sciences, social thought under the regime of social sciences still consists of thinking about secluded island of national socials, presupposing that the social within these national social islands could be understood by confining social thought to reflecting on nation state socials. Social sciences have more or less no clue about any social beyond the borders of their nation states, not to mention any insights about how the global battles about political and economic power craft the entire social life within all those seemingly secluded social entities as a means for these very battles. Accusing them that they are ignorant about other state social is not only downplaying that social science thinking does not care about socials beyond any nation states, it misunderstands that thinking about national social is the natural unit of analysis in which social sciences think, thanks to their illusion about nation state socials as a secluded entity, in which their national social could be understood.

Despite of the fact that the very whole post world war II world shares essentially the same society system, the capitalist economy and the—US—concept of nation states all using their individual state social for their global business and policy affairs, global social thought under the regime of social science thinking does not want to think about the social as a world's nation state social, but is—still caught be the sovereignty of nation states—committed to the idea of the reign of parochial thought created in and about secluded islands of knowledge, all creating their island-like theories. And, if they do deal with any other island-like social, mainly comparing nationally constructed theories, they are seriously challenged if social science thinking crosses the borders of any nation state social.

...by assembling theories about nation state social biotopes....

Global social thought in the social sciences that detects the world's social and that crosses the borders of the national social is the assemblage of the secluded knowledge about nation state socials.

If social science thinking crosses the borders of its national social biotopes—it continues to look at the world's social as an agglomeration of nation state social theories and becomes "inter-national" by comparing their nationally constructed thought, theories created from the very state science thinking view on the social within their state biotopes.

What elsewhere would be considered as violating the most fundamental rules of social sciences theorizing and rejected also within the social sciences as nationally "biased" thought, thinking in national "perspectives" is ordinary practice in international social science activities. The national social is not only the unit of analysis but an explanatory framework through which social science thinking theorizes about the national social. Presenting social thought under headlines like ".....from a Chinese perspective", are not rejected as obviously biased knowledge, but very welcome as enriching the assemblage of theories, not only constructed about nationally confined knowledge, but knowledge constructed through the pre-supposed thinking of a nationally biased view about any topics.

Assembling knowledge by preferably carrying out and comparing country studies, inter-national theorizing in the social sciences, consists of additive knowledge about multiple nation state socials that is lacking any commensurability. Since such knowledge assemblage compares nation state social without knowing any tertium comparison is the nation state socials share and against which they could be compared, the result of these studies is to detect a never ending round-about of non-understood divergences. How could they? Since social sciences only know how to think about the individual nation states social, they have no concepts of what a nation state essentially is, and are thus unable to identify and distinguish what nation states and national societies across the world share and what not.

As a result, thinking in nation state "perspectives" introduces any national, mostly historical peculiarities of nation states, as an imperative theoretical means needed to theorize about the nation state socials—and discloses the extent to which international social science theorizing drowns theorizing in the monstrous cognitive circle, that provides to share the nationally peculiar constructs and categories, the national "perspectives" as a pre-conditional means to understand them. To give just one example of this dead end road thinking in such international comparative country studies:

"These difficulties are not only due to the difference between English and French. They probably also reflect the French conception of knowledge, which puts an emphasis on explicit and scientific knowledge, and the French conception of learning, which traditionally puts the emphasis on formal education and training."[11]

Since social science have no clue about what the essentials of a nation state social is and, hence, have no categories theorizing about a nation state social, they cannot distinguish between any essential of a nation state social as such and their historic peculiarities. Hence, social science thinking considers any social phenomenon in any individual nation state social as a unique phenomenon of any individual nation state social.

Thus, any general features of the nature of humans, essentials of the construct of nation states or historical peculiarities of a particular nation state are undistinguishable for social science thinking. Hence, social science thinking not only knows things like a *"French conception of knowledge"*. Nation states undoubtedly craft the living conditions and the life of humans and do this to an extend that made Marx talk about his notion of a "Charaktermaske", critiquing that the most liberate inhabitants of the nation state societies without having a clue about this only execute what they are forced to do by law and consider this as only executing their most individual peculiar views and life agendas Thus, do the social sciences, when they assemble knowledge about nation state socials and when they compare them, identifying the historical peculiarities of their nation state social with what their nation state social is: Unlike China, France is the French "manifestation" of the French nation state

Undoubtedly, humans have created different concepts of what they consider as knowledge. However, imagining a concept of knowledge, that defines a *nationally* peculiar mode to construct thinking, a national concept of what is human's nature, can only be imagined by thinkers for whom the nation state is the almighty power even able to implant a nation state view on humans, here on how humans think, as a second, quasi national human nature.

Once any national peculiarities are identified as the particular nature of a nation state social, for social science theorizing looking

---

[11] Mehault, P (2007), Knowledge Economy, Learning Society and Lifelong Learning—A Review of the French Literature, in: Kuhn, M., *New Society Models for a New Millenium,* New York, Peter Lang, p 67

beyond the borders of their national social requires to share these nationally unique concepts as a precondition to understand them in the comparative view on the world's national socials. Not surprisingly these studies ever end up in the complaints among all the inter-nationally thinking social scientists, that the others are never understood by the others.

Theorizing in national perspectives and assembling such nationally constructed knowledge, is the only way social science know to creating social thought namely in the rightly called international science encounters, that has indeed so much internalized the constructs of state constructed societies, that the naturalization of these state constructs only allow them to recognize the national peculiarities, the historically particular interpretations of these constructs as the essentials of the individual nation states and, hence lead to a new version of globalized ignorance among social sciences about the other national socials.

As in our example about an international comparative view on the sphere of education, social science thinking is not able to see what this particular national systems essentially shares with the education systems of the countries against which it is compared, but, falsely—locked in their thinking in comparative national "perspectives"—identifies the particularism of the national interpretation of the state education system, here the French education with the nature of education in France, what is only the peculiar variation of the way to interpret essentially the same education system, the French education shares with the education systems against which it is compared. Excluding the systemic fundaments from reflecting about education in thinking in national "perspectives", results in considering the peculiarities of the nation state constructs as their essentials and creates the, indeed, very national view not only this French scholar advocates as the key to understand education in this country and across the world's national state socials.[12]

---

[12] It is exactly this, describing the phenomenology of things, identifying the nation state fabrication of humans as their nature, rejecting any insights into what is manifested, insisting on the "uniqueness of the reality", voicing what he observes as theorizing about the social what Weber's phrases as the way of thinking in what he therefore calls "Wirklichkeitswissenschaft, when he says: "The type of social science in which we are interested, is an empirical science of concrete reality (Wirklichkeitswissenschaft). Our aim is the understanding of the characteristic uniqueness of the reality in which we move. We wish to

The global indifference among social sciences about the other nation state social is thus the inevitable consequence of theorizing about the world's social through thinking in national "perspectives" about the national socials, the particularisms of nationally constructed categories, presenting a nationally peculiar concept of humans as the essential of a nationally constructed human nature—*the* elementary "enlightened" form of a theoretical racism in social science thinking.

...off-thinking the world's social...

Constructing theories that present the social as secluded national entities, and consequently, as in our example, presenting the national fabrication of humans as the nature of humans, is a construct of the social sciences in the imperial nation states and the claim to international social sciences, internationalizes the social sciences approach to global social thought that is that the world's social must be reflected on as accumulating such nationally constructed theories about secluded national socials, interpreted through the national peculiarities constituting the unique national "perspectives".

Presupposing the national socials in developing countries as such social entities secluded from the world is most obviously almost impossible, since it is too obvious that their national socials are a product of the imperial nation states. Societies that only exist as a means to serve the economic and political power needs of the imperial countries could hardly create social thought about their national socials that presents the image of national socials and of the world consisting of such social biotopes, the social sciences in the imperial world present as the theoretical entities through which theorizing about the social must and could only be understood.

However, also within these imperial nation states socials, thinking about the national social as secluded from the world's social

---

understand on the one hand the relationship and the cultural significance of individual events in their contemporary manifestations, and on the other the causes of being historically so and not otherwise. " Weber, Max (1949), *The Methodology of Social Science,* translated by Edward A. Shils& Henry A. Finch, Glencoe Illinois, Free Press 1949, p 72 implies to off-think the impact the world' socials have on each other via their nation states and via their economies.

Social science theorizing in the imperial countries does, in fact, precisely this, theorizing about the world's social as theorizing about secluded social biotopes, unaffected by each other. Theorizing in globalizing social science thinking is in the first place to off-think the world beyond their national biotopes.

A few examples may show that this presupposition, thinking national socials as socials secluded from other national socials, requires to practice thinking as the determined ignorance even about what social science surely do know about how the world beyond the individual national biotopes affects the national socials.

That thinking about the "happiness" of people—let aside what this dubious category ever means—for social science theorizing must be as any other phenomenon of global theorizing considered as an issue related to nation states constitutes for social science thinkers the nation state as their comparative unit of analysis, to find out in which country one can find "differences in happiness":

> "This item response theory methodology is first applied to assess the differences in happiness across selected European states." [13]

Admittedly, theorizing about the happiness of people is certainly quite an odd topic for social sciences and has the strong taste of EU-propaganda, comparing happiness across European nation states, nation states, which day by day boast with their agendas making Europeans an attractive "human recourse" and thus Europe an attractive global business location.

However, it is not the odd topic and the propaganda mission of such studies, but what is important is that this way of thinking is a most typical example for the globalized way of social science thinking, may this be about the happiness of European humans. It is in fact very typical for global social science thinking that thinking about happiness must be thinking about the *"happiness across selected European states"* and, thus, must be a matter of comparing nationally constructed humans and the differences of their happiness a matter of nationally constructed data, "indicating" how they feel as nationals, as citizens of each country. Thinking about na-

---

[13] Rynko, Maja, *On the Measurement of Welfare, Happiness and Inequality*, European University Institut, http://cadmus.eui.eu/handle/1814/20694

tional socials, dividing people into different national socials and, hence, off-thinking any other national socials of the same shared political body as separate social entities while theorizing about a group of national social though all strongly politically and economically bound to each other in the same "European states", is a masterpiece of social science thinking. It is a masterpiece of ignorance, to off-think the very relatedness of these nation state socials, thinking them as nationals, though they are all made the same European socials. Off-thinking the other national social while comparing them, even when they are subjects of the same political entity illustrates a typical method of global social science theorizing off-thinking the world while thinking about the world.

Especially within a group of countries, where the social life of its citizens is so much a product of the all kind of carefully administrated interactions of the nation state socials ruled by a joint currency and a supra national governing body, only a view that does not know any other access to thinking about the social than the presupposition of the social sciences, that any social must be understand as a national social, can present the happiness of citizens as nationally constructed features of humans, off-thinking that the social reality of EU-citizens is more than in many other national socials, a product of the interaction of the nation state socials within the shared political body of the European Union. Even in a case, where the social reality is so obviously *made* by the interactions of nation states and an inter-national, here the European economy, social science theorizing must think about the social as carefully secluded national social biotopes, off-thinking that it is only the interrelations between the national subjects which craft their life, may it be their "happiness".

Presupposing that the world's social must be explained as theorizing about secluded nation state socials and the determined ignorance it needs for social sciences to think about any social phenomenon as a nationally constructed phenomenon, might be illustrated by another example, about a certainly less dubious topic, the famous study by Bourdieu about nothing less serious but a study about the academia.[14]

The fact that the title of the famous book "Homo Academicus" insinuates that Bourdieu—a social scientists, who certainly cannot

---

[14] Bourdieu, P, (1988) *Homo Academicus,* Cambridge, Polity Press

be suspected as being an a priory nationalist thinker—presents under this title his thought about a particular human species one can find across the world, the academic, obviously does not irritate this prominent social science thinker, that this book presents under this most general title a study about two faculties at one university in one country, France, not to say anything about who the homo academicus is, but to theorize under this very title about the question, why academics in those two faculties, one standing for conservative and the other one for progressive academics, joined or did not join the protest movements in France in the 1970s.

As if there was no such academic species beyond France, thinking about a university in France under a title that discusses a relevant species of nation state humans anywhere in the world, a species one can find in any of those national biotopes across the world, it does not violate social thought in social science thinking to think under this title about the members of a national university, since social sciences thinking is thinking in national entities. It simply does not come to a social science thinkers mind that looking at the "homo academicus" beyond a singular national social, might not only be enlightening to study this typical nation state creature and, by doing this, to justify the very general claim of this book title, the "Homo Academicus". If social science thinking creates social thought it is social thought that identifies any human or human activities with their nationally confined existence and it does not come to a social science mind to consider reflecting about the phenomena outside or even beyond their individual state constructed socials islands, because they are obviously not able to see any nation state construct of humans other than as their nature—just as if the homo academicus was a mere variation of the very same human mankind.

Scrutinizing the question, why different departments in the Sorbonne university have different political positions about what moved the student movements in France, somebody like Bourdieu knows of course very well, that this protest of—firstly—many intellectuals was massively caused by their opposition against the US war in Vietnam, an object of concern that, he though very well knows, cannot be found in his study about the Sorbonne life of faculties in Paris. This is about world politics and, thus, not a matter of sociological thinking. And Bourdieu, a social scientist, a sociologist, a leftist thinker, what is he doing to answer his question, what made academics join the protest about a war or not, a war in which France was massively involved as the former colonial power? He digs into tons of data which all have absolutely nothing to do with this war and, instead, he searches in all sorts of data that deal with the most ordinary national categories, any sociological studies finds enlightening, when they investigate whatever sociologist find exciting to know about, seriously believing he could find the reasons for the different political positions in the different departments of sociology and the law faculties in things, sociologist make responsible for whatever they research, may this be to find out why people demonstrate against a war anywhere in the world, why they drink Coca Cola or why they beat up their children: Social science thinkers, here theorizing about academics, know where to find the answers for anything and dig into data about things such as: their family status, their sex, their age, the newspapers they read, the cars they drive, the districts in which they live, the size of their apartment, the number of children, the age when they marry, etc. etc., in short, all the utterly nationally constructed features all serving as "indicators" for a sociological thinker showing to a sociologist the extent to which citizens accommodate themselves to what sociologists have defined as their ordinary national "roles", ever—strictly tautologically—carefully observing their one and only concern if any citizens deviates from their roles that build their community. For Bourdieu answering his question about who joins the protest is as easy to be answered as any of these sociologically constructed tautologies is: People who conserve the roles, live in posh apartments in Bois de Boulogne, etc. etc., indicators they were attributed, are conservatives and—smartly concluded—do not join the protests; and, people, who do not marry when they should, drive a 2CV, live in cheap flats in Montmatre are those who deviate from their roles and—join the protest. The obvious nonsense of this way of proving any theories about what ever, is not the point here, the point is here how this nonsense manages to pull the issue of a protest against a *war* towards what sociological thinkers do with any topic, that is that any sociological explanation, even for those phenomena which most obviously are motivated by anything beyond the inside of nation states, are torn towards reflections about the interrelations between individuals and the society, as sociologists like to problematize their national social.

In social science thinking a world's social *made* by a multiplicity of nation states crafting social life within these national biotopes to function for the nation state's world social does simply not exist. Though nation states and their interactions are the most powerful factor in the globe shaping the life of humans within their territories via and for the global actions more than any other social subject could do, last but not least by wars, social science thinking reflects on the life of humans within these national entities unaffected by any outside world. When people in a country like France articulate their protest against an imperial war against people in a country they were involved in as former colonizers, social science thinkers redefine this political view on world politics towards the question about the extent to which people are integrated in the national social biotope or not and present this to explain why people stand up to protest, while others stay at home. This, thinking about the social and thus the world's social as thinking about the concerns of any relatedness inside national biotopes is the only way social sciences are able to even theorize about why people protest against imperial wars.

As if the only resources nation states use for their global interactions was not their command over the recourses humans and nature provide, as if what they do in their global political actions with these recourses and how these global actions affect people's lives and the nature, the imperial social is simply no topic in social science thinking and if the outside world occurs, social science thinking off-thinks the world towards a matter of any inside nation state social, here the relation between the nation state and its inhabitants.

If at all, social sciences look beyond their individual national entities and reflect on them via a sub-section of political and economic theories, they look at it for the cultural entertainments of the well-educated via mainly historical knowledge in some orchid humanities and via the racist exoticisms of Anthropology, more recently replaced via the methodological exoticism of cultural studies. However, the major social science thinking does not want to notice that it is the interactions between the national entities that shape the live of all their idealized human creatures within their biotopes they reflect on, ever untouched by the world.

Admittedly, one may very well take any social phenomenon within the national socials as the point of departure for theorizing about it. However, only social science theorizing remains biotopian social thought to which the inclinations of the national social phenomena in a world ruled by nation states will not occur. Thinking that may well start from any national social, will sooner or later encounter that the nationally confined appearance of social phenomena is an illusion that originates from an idealized perception of the sovereignty of the national entities. The fact that the foreign affairs of nation states are not the business of its citizens but of politicians does not mean that they do not matter for them and any theorizing that may very well depart from the nationally constructed social, thinking about what crafts national humans within these national social bodies, inevitably arrives at detecting the illusion the sovereignty of nation states insinuates, that social life could be understood within the national social constructs.

Humans newly born, entering the world's social without ever been asked enter the world's social defined as citizens, before they can think or decide, they are made in an act of demarcation of one nation state against the rest of nation states a member of a nation state. Attributing them the status of the citizen of a nation state, defines them as a humans, subordinated under the governance of an individual nation state, excluding other nation states from any command over a human on which a nation state has declared his reign and decides with this demarcation about the main living conditions of humans, once they are made citizens of nation states. From there on humans are made a human, constructed as nation state creatures, constructed with the totalitarian claim making thus humans nothing but a committed national creature, sharing the rationales of a nation state as the basis of his life, an act that takes for granted that these nation state citizens carry the past and the future missions of the nation state, citizens have been made a member of.

Which nationality a human is attributed, may this be in an imperial nation state or in a nation state of the developing world, decides about more than only the living conditions of humans. With the attribution of being defined as a member of a particular nation state the life agenda of humans is substantially pre-configured by the nation state rationale that declares their new members as members of a nationally ruled population, a means in the battles with the global rationales of other nation states. Becoming a citizens of a nation state claims its citizens as being a lifelong belongingness, as if humans were owned by the nation, defining this citizens as the exclusive citizen of a nation state, a lifelong claimed element of the governed human substance of this nation state, in the worst case a "natural" worrier in the wars between nation states about global political and economic power. And this worst case is not at all such an exception, considering that the world *is* a world of war, of wealth and poverty. It is for this very reason, that those humans, who are not involved in wars among nation states, are in fact considered as historical exceptions.

That a human defined as a citizen of a nation state serves the nation state that has defined him as his citizens, that a citizen serves the rationale of this nation state is taken for granted. Within this rationale the nation state human is entirely free to choose among the option of life agendas, nation state societies offer their members. Where and how this human spends his life along the life options nation states offer, is also taken for granted, trusting that the totality of the dependency of any life options on the nation state constructs ruling any of these options, makes citizens a committed member of a nation state social. Imposing the nation state rationale into the citizens' life via a totalitarianism of laws that regulate any aspect of human life, whatever life option citizens chose, force their live agenda towards coinciding with the nation state rationales and thus make them an element of the global political agenda in the global political and economic battles among nation states.

In nation state ruled societies, in which striving for any life objective is dependent on the disposal of money and in in which the post war rationale of nation states makes the value of any national currency the subject of the value judgments of the global financial market, whatever people can do or not in these nation states, it is a most dependent variable of the global financial market and their bargain with national currencies. A nation state society model, making humans private property owners measuring the properties in national currencies and at the same time make the decision about what the value of these national currencies is, a matter of the global financial market, makes what these property owners own and hence, what their life agenda is, a dependent variable of the global economic business of the global capital. Presenting national currencies as a service for the inhabitants of a national economy to supply them with goods is an odd illusion of economics about a neat life in a nation state biotope.

The majority of people in all nation states around the world with their capitalist economy are dependent of gaining there income by selling their labor. The availability of jobs depends on the attractiveness of the country for the global capital and the price for labor is—only—one factor for their decision, in which nation state they invest their money and, as capitalist and politicians phrase it, in which they provide jobs.

The whole working population is an economic global hostage of the global capital, since nothing less but the sheer existence of the world's people depends on the growth of wealth of the few people, owning enough of the world's wealth to make the world's nation state creatures a means of their business—or not, and—if it serves increasing their wealth, to employ the world people seeking an income, depends on their decision, in which nation state the global capital invests its money. Making the exploitation of labor more attractive for investors than other nation states do, is therefore the concern of nation state policies competing with others of the same kind to attract global capital and it is this global competition making the main parts of a national population an attractive workforce that mainly crafts the economic fundaments of the living conditions of the world's nation state inhabitants in societies, in which realizing any individual life agenda has been made dependent on the availability of money, they can only gain via offering labor to the global balance pros and cons of the global business world.

And what are social sciences thinking while theorizing about these nationally constructed global creatures, the citizens of nation states? They off-think the global rationale that shapes them as national creatures and present them as merely nationally constructed, off-think the world as the world of nation states, into which they are thrown as national creatures, by interpreting all the national constructs, may this be the nation state construct of an individual, the nation state constructs of a family, of a student, a worker, a capitalist, a retired person, a teacher, gender, a childhood or as a politician, making all these nation state constructs natural constructs originating from the human nature, and the nation states, abstracted from their political agendas transformed into political bodies, ever responding to deficiencies of the human's nature constructs, disarraying them from their nation state nature and thus from their nature as a means for the world's nation state affairs.

To illustrate this social science manner, off-thinking the imperial nature of nation states, crafting the live of the inhabitants of nation states, only with very few examples along some major cornerstones of citizens' life:

In the field of education, that is seemingly far away from being affected by the interrelations of nation states, it does not need Pisa studies or the like to notice that presenting education as a service of the national education system for young people to develop their talents is a neat idea of an educational ideal, but in reality rather about the development of "human resources" of an educated workforce, mainly the imperial countries consider as a major weapon in the global competition about benefitting from the growth of capital offering of a well-educated workforce for attracting capital towards their territories. Off-thinking the imperial nature of education educational theories discuss education as the challenge to serve and develop life perspectives coinciding with the talents of humans and the needs of a labor market.

The salaries these people get for their labor are negotiated with their employers and are presented by social sciences as an income that depends on the efficiency of their work and their levels of skills. However, how much they ever work, how much they can buy for what they get not only depends on the exchange rate of the currencies in which they are paid and the prices for all the imported and exported goods; if work is at all available on the national labor market depends on consideration of global investors, in which the national labor market is only one factor in a comparison with the work forces offered on other national labor markets.

Arriving with their talents in the world of work, sociologists detect that the real wealth of people, who struggle to earn their income via jobs, is the "social capital" especially the have-nots own, the currency of the poor socials ever discussed as the real wealth of belonging to a community, not to a particular nation state, no, sociologists are more promoting the much more abstract idea of being part of a social as such, of what Bourdieu calls "durable networks", which is much more valuable than the ephemeral value of real capital and thus compliment such theories detecting citizenship as the real value with investigations about their happiness—carefully distinguished along nation states. The world's social, crafting peoples live has been dissolved into the sociological value of their belongingness to a nest, any community, may this be a family, company, or a football club, in short any of the social institutional constructs in which nation states organize the lives of their inhabitants. The world's social—thought-off by the social sciences.

In short, there is not any sphere of life that nation states do not arrange with their power to set the rules for living that is not ruled by the concerns of nation states about their power position in the world and for making their territories and their people an appealing offer for global investors. And there is no social science thought that does not off-think the global affairs of nation states and interpret the impositions of the global agenda of nation states on the citizens as a service for their life agendas, at least those social sciences interpret into what humans are aiming at.

Reading the sovereignty of nation states over their territories and people as if the national social was not the result of the impact of the relation of nation state on the social can only occur in imperial nation states which are able to dictate these relations, which does though not mean that they are not affected by these interrelations.

While in most nation states on the globe the illusion that the individual nation state is the locally confined nest of their citizens can hardly occur, since already the existence of the inhabitants of these nation states at the wrong time on the wrong place might be considered as rather disturbing the nation states agenda, it is a false conclusion in those few nation states, which rule the world of nation states, to conclude from the fact that these nation states are able to shape the living conditions of their citizens, since they dominate the world of nation states, that the life of these citizens is not shaped by the whole of the nation states.

Off-thinking the fact that the imperial nation states use the people and the nature of the countries they dominate, by interpreting the social in the imperial countries as unaffected by the relation of nation states, is the cynicism of social science theorizing about the national socials as secluded biotopes also in the imperial countries.

Social science thinking is not only ignorant about the world and rather off-thinks the world's social, omitting the global empire of a nation state ruled world from social thought.[15] As a result social science knowledge consists of secluded islands of knowledge, all looking at the social live of nation state socials, abstracted from the world of nation states, which craft the national socials across the world more than any other social actor—an imaginary way of constructing the world's nation state social preferably practiced in the social sciences of the imperial world.

Notwithstanding, there are a few social scientists, preferably economists, who rescue the scientific reputation of the social sciences and in fact theorize about a social across nation state socials.

> In a recent book (2012), Joseph Stiglitz, a former Nobel Prize winner in Economics argues that rising income inequality is one of the main factors underlying the economic and financial crisis in the United States.... The social and economic challenges associated with rising income inequalities have gained prominence in the public debate, after the publication in 2009, of a widely cited book by Richard Wilkinson and Kate Pickett entitled "The Spirit Level, Why More Equal Societies Almost Always Do Better". Using cross-national data, the authors show that income inequality correlates with lower levels of social capital as well as with a host of other social challenges from poor health, crime, to underage pregnancies. The current report takes part in this debate by examining the bivariate correlations at subnational level (NUTS 1 level) between income inequality and indicators of education, health, criminality, political participation, social capital and happiness at the EU level."[16]

Not surprisingly the social scientist rescuing the scientific honor of the social sciences is a Nobel prize winner, who "argues that rising income inequality is one of the main factors underlying the economic and financial crisis in the United States" ... The insight, that poverty is not a crisis, for those who are poor, but a "main factor underlying the economic and financial crisis in the United States", inspired the before mentioned study about the happiness of EU citizens, to "*take(s) part in this debate by examining the bivariate correlations at subnational level between income inequality and indicators of education, health, criminality, political participation, social capital and happiness at the EU level.*" In other words, this

---

[15] The little attention very rare studies on the global nation state world gain by the social science discourses, e.g. the one by I. Wallerstein *"The Modern World System I-IV"*, New York 1974-1989, illustrates that social science thinking does not know how to deal with global theories and therefore treats them as more exotic intellectual enterprises.

[16] Rynko, Maja, *On the Measurement of Welfare, Happiness and Inequality*, European University Institut, http://cadmus.eui.eu/handle/1814/20694

is a case, in which social sciences theoretically do cross the borders and think beyond the national biotopes. Do they?

The EU political elite and their economic global players "at the EU level" will shiver if they are told that poverty and happiness are bivariately correlated, spiced with a pinch of criminality—"at the EU level", a notion that reveals that social sciences are even lacking a decent vocabulary for naming international social entities that are not nationally contoured, they try to circumscribe as what? The interplay of nation state, here the EU, more than in many other parts of the world crafting with their battles about power the lives of their citizens in the EU and beyond—is a "level"—an additum to nation states, off-thinking their imperial alliance, imperial towards the members of this alliance as to the world's nation states and their citizens—except that their happiness must be correlating with their nation states.

... reflected on through the nation state constructs

Social science thinking compliments the secluding social thought about national knowledge islands, with a way of thinking about the national biotopes that reflects on the national biotopes through the—idealized—perspective of a nation state rationale and the rationales of the nation state creatures.

Thinking as reproducing the "*concrete reality*" *(Weber)* is an approach to social thought, that not only creates the national biotopes as—to phrase it in the language of social science thinking—their "unit of analysis", but the nation state constructs as the "analytical perspective", through which social sciences theorize about the national biotopes. Social science thinking thinks about humans within these secluded biotopes through the perspectives of the constructs the nation state social has attributed to their lives interpreting these constructs as if they were their nature.

Social science thinking about the national social biotopes is thinking through the view of subjects, cognitively reproducing the practical views these nation state social constructs define as who these humans are and how to reflect on these subjects: Children, students, retired people, employees, tax payers, employers, politicians, families, all these creatures which are nation state definitions, social thinking does not want to scrutinize, but instead takes these nation state subjects and the practical concerns they have as these nation state subjects about the social reality as the volontary perspective through which they create social thought about the national biotopes.

To give just one most ordinary example from social science thinking, how it re-constructs humans as nation subjects and how it re-constructs the concerns only these nation state constructs have as the perspective through which they theorize about them:

> "If skill requirements increase, low skilled workers will be under increasing pressure, in the industrial sector and in some service sector. Demographic evolutions could reinforce this tendency." [17]

As for any social sciences theories the subject in this little scenario is a typical nation state creature, a "low skilled worker", a most typical human as they usually appear in social science theorizing as the perspective through which they interpret social their life: Their actions are ever *coping* with the life definitions they have been attached as nation state figures crafting the social thought of the social scientist ever *caring* about these state creatures. Knowledge these creatures have are defined as "skills" and "skills" for these creatures are not an ability a subject owns, but a foreign request these subjects must ever obey; increasing skills are never increasing the means of these subjects to increasingly better know how to do something better for their purposes, but a *pressure* to *subordinate* the subject under ever alien *demands* ever coming from any unknown subject, developments of ever *natural necessities* which ever appear like natural requests, subjectless requests, this typical social science subject, the "low skilled worker", has no influence on, he only can *obey to avoid anything worse*. "Demographic evolutions", again a view of nation states on their "populations", accordingly "age" and "aging" are not natural aspects of human life but social science categories which reveal a concept of time that presents the life time as a dooming threat ever thinking the future of all their creatures as oncoming additional threats, reinforcing all the coercions humans must obey—just as if it was the nature of human life to be an inevitable natural threat.

In short: The image of a typical nation state human, a citizen, a self-domesticated creature striving for his life aims by crafting his life as an adjustment to foreign requests, life and time as threats

---

[17] Mehault, P ....p 80

requiring obedience, presented by social the sciences as necessities inherent in the nature of the social world, a world of threats presented as the nature of social life, threats, established by unknown subjects which never appear, ever commanding social subjects to follow them as if they were factual necessities.

This is the human, in all its facets a construct of nation state, a typical citizen, social science thinking not only reproduces in their way of theorizing about the social, but it is the very view these nation state creatures have on life as a life of threads, interpreting the social life, its agendas and biographies only nation state constructs have, as factual necessities, social science theorizing shares as the perspective for theorizing on this nation state *made* social, just as if the nation state view on the citizens was the same as the view of the citizens. Off-thinking all the conflicting interest between the nation state and the citizens and presenting the nation state mission as serving to solve the citizen's "problems", they only have due to their nature as nation state constructs, is the way social sciences reflect on the nation state socials.

However, social sciences are not thinking nationalistically, their reflexive imprisonment in the nation state constructs ever interpreted through the view of the nation state on his citizens, presented as the view of the citizen's concerns, is much more fundamental and subtle. Their perspective is not the perspective of any real political agenda of any particular, real nation state, but the view of the nation state and the nation state's humans systemic constructs, the view on humans they construct and reflect on through the perspective of these nation state creatures, the naturalization of the state into humans nature, the politically defined citizen, that builds the basis for all their thought about the inhabitants of the secluded biotopes.

Only social science thinking is not able to distinguish between humans and citizens, life objectives of humans and the ways state laws domesticate how and to which end they must be pursued. Social science thinking cannot see that there is not only a difference between humans and their life objectives and their lawful definitions; identifying both they have no headache to interpret their restriction and their channeling towards nation state objectives through their lawful definitions as a necessity of any social life and thus the domesticating and channeling of the state definitions of the humans life objectives as a help to achieve these life objectives of humans. Thus, they detect the nation state society as the best possible natural home for humans and the state view on these humans as the view of social thought about humans.

Though no human ever demands work, however, in social science thinking, it is a service of state policies to provide people with jobs and the difference between the demand for a salary paid for labor, is for social science thinking an ignorable feature, since the state law decided by all its force, that there is no life existence without offering labor on the labor market to those who pay for labor for those who have no other resource but selling their labor. Thus, social science thinking identifies the demand for money with the demand for labor, since it considers the state regulations for pursuing people's life aims nation state policies set up for humans life as life options under which they are only allowed to pursue them as the natural living conditions of humans.

Consequently, humans who are existentially dependent on paid labor, but who are not offered getting money for work, are defined as, not lacking money but lacking labor, they are therefore defined by the nation state as "unemployed" and state science thinking not only shares this state view also defining people lacking income as lacking employment. Caught by the idealization of the nation state as the community of humans, social science thinking goes a step further and interprets the concern of having no income from paid labor as a the "risk" to be no longer a member of this community, a risk they therefore call "social exclusion"—just as if being defined as seeking labor was not the way "the community" secures that nobody can escape from the economic logics of the communities' labor market, also when it fails to provide the "job-seekers" with jobs, who only seek money.

Humans are getting older—more precisely, some humans in the imperial nation states, while in others parts of the world rather the opposite is the case. Nevertheless, for social science thinking this is a—problem. Dying later a problem? A problem for whom? People who are living longer find themselves in social science thinking through nation state perspectives in what these social sciences regurgitating what the political discourses define as their view, as an "aging population" causing all kind of problems—as usual in social science thinking, problems for the nation state and its social policy agenda.

Founded on the naturalization of nation states creatures and the nation state as a community aiming at serving its inhabitants, social science thinking develops its affinity to the rationales of nation states towards nationally driven reflections, ever critically observing if and how the nation states moves towards its idealized objectives. Social science thinkers are no political nationalists but passionate devotees of their ideals, ever critically measuring nation states against the ideals they create and by doing so develop a critical affinity to nation states agendas.[18]

...ever critically measured against idealized nation state rationales

To stress this again: Probably with the exception of some branches of the political sciences, that merely reflect on the functioning citoyen and the functioning of the polis, measured against national policy priorities, thus a theory which can be hardly distinguished from the ways political practitioners think, all other social sciences theories are no nationalist nation state thought, no political party supporters of a nation state or even of any nation state policy agendas.

However, thinking through the nation state rational or, better, an idealized nation state rational, though also not thought through the concerns of a real policy agenda of the political elites, but through the idealized agenda of particular nation states, in *global* social thought in the social science approach, it is seemingly not considered as violating even the scientific dogmas of the very social sciences, they elsewhere would reject as prejudiced thinking, if global social sciences theorizing contributes social thought explicitly theorizing through national views on their nationally confined social. In global social thought under the regime of the social sciences prejudiced thinking through "national perspectives" is rather seen as a necessity of global thinking consisting of an agglomeration of such nationally constructed theories. Thinking through the particularisms of nation state constructs in global social thought is

---

[18] Thus, such as the categories "unemployment", "social exclusion" and "aging populations" mirror, thinking along the idealized rationales of nation state and their social constructs materializes in the particularity of scientific categories of nation state languages.

a normal must in social science theorizing. Phrasing thought as seen through any "Chinese" or whatever national perspective is a most ordinary contribution to global social thought under the social sciences.

Nonetheless, also this very explicit presupposed thinking through the national perspective of individual nation states in global social science thinking does rarely advocate any national policy agendas, but rather most critically measures the political practices against the ideal it has about nation state and detects all kind of violations of these nation state ideals in the political practices of real nation policies considered. What social science thinkers and nation states share through their idealistic social thoughts are their ideals of national policies, policy practitioners only use for the political propaganda, not to set up their policy agendas, a presentation of politics for which social sciences provide with their criticism the ideological interpretations of these political practices and the objectives they supposedly aim at and are therefore social science thought that is though rarely appreciated or rewarded by the political elite, which rather sees social science knowledge most critically, especially if social sciences insist on their nation state ideals against the nation state practices.

Social science thinking not only incorporates an idealized state mission when reflecting on the social from an imagined practical view the nation state creatures in the state islands have on the national social. Social science theorizing knows much most subtle ways of incorporating an idealized mission of nation states, only social sciences make up in the construction of social thought, ever off-thinking the imperial world's social as a precondition to induce their nation state ideals into their theories.

To illustrate this with an example: The Frankfurter Schule, namely Adorno, can certainly not be suspected sharing any nationalist ideas and in Adorno's case, the least one could assume is that Adorno thinks as a German nationalist.

Adornos reflections about what he circumscribes as "Auschwitz" might be an example for a discreet way to theoretically appropriate the individual state rationales in social science thinking, presenting activities in a war as an issue confined to the concerns of a nationally secluded social, presenting war actions as secluded from the world's social, and at the same time most critically opposing nation state practices with an ideal mission of a nation state:

"Die Forderung, daß Auschwitz nicht noch einmal sei, ist die allererste an Erziehung. Sie geht so sehr jeglicher anderen voran, daß ich weder glaube, sie begründen zu müssen noch zu sollen." [19]
(Postulating, that there may be no other Auschwitz again, is in the first place one towards Education. This is so much a primary postulation, that I believe, I neither need nor should I motivate why.") (Own translation MK)

Making Auschwitz a matter of education, and making this so determined a matter of education, that any reasoning, why avoiding another Auschwitz must be seen as a matter of education, would violate the high moral mission education is attributed by this thinker; making Auschwitz a matter of education, does, however, only not need to be "motivated" only for a mind, that considers Auschwitz as a matter of the—failed—morality of people, the moral responsibility and the moral failure of badly educated Germans, thus extinguishing any reflections on the political rationale of the German nation state war activities, and that thereby discloses the very critical and very German, very national moral mission, this philosopher considers as the mission the nation state he reflects on as a social science thinker should aim at in the future.

And this is a most critical mission of a most critical social science thinker that, thanks to his interpretation of Auschwitz as a moral issue that does not want to know anything about any real national reasons for Auschwitz, though somewhere matches with *the very real* post war policy rationale of German imperial politics after World War II until today. By making Auschwitz a matter of moral education, rather than a matter of the cynical rationale of wars not only world war II practiced—this critique dissolves the cynical rationale of a war, that eliminates not only the enemies of the country against which it carries out war, but also any—may they be imagined or not—enemies inside, ignoring any political rationales of the war this critique dissolves not only the German post world war II policy mission of the German political elite into a mission of moral education, the German war as the German failed moral mission, by making the rationale of a war and the lost war a matter of a lost moral mission, this *critique* of the German nation state as a moral failure manages to off-think, what only the most violent imperial actions of nation state can do by transforming the

---

19   Adorno, T.W., (1971) *Erziehung zur Mündigkeit*. Frankfurt a.M., Suhrkamp, 1. Auflage, p 88

rational of a war into a failure of his responsibility for educating the values of humans morality. Shifting the debate about Auschwitz from the debate about the political rationale of the German political elite towards a moral failure of the German nation state and from there as a moral failure of *the* Germans, is to make this issue, firstly a matter of morality and via doing this, secondly a matter of *the* German as a member of this nation state, a subject that is attributed the policy rationale of the nation state as if war was the committed rationale of their citizens, not willing to distinguish between the citizens and the political elites, those citizens nation states, as the German nation state, instrumentalises for the rationale of an imperial war, all unified in social science thinking in the shared failure of a joint failed moral mission of the failed national "we"—towards a new purified national "good" moral we.

Thus, creating with the self-critical confession of being a moral failure, creating *the post war national good "we"*, a morally shared guiltiness of *the* German, unifying all Germans in sharing a now thanks to this confession national cleaned German, by committing the German to a joint moral failure, is *the* post war concept of the German nation state, which very much helped to build a new German "we" the "we" of guilty people, a new German national identity, after the old one, the image of the higher race, did no longer work due to the lost war. Very soon after, swearing "Never again war", this new self-incriminating "we" re-established the new German nation state and the military forces—preparing for the next war, against the same enemy. Confessing a national moral guiltiness of the German, nobody is allowed to question, if he does not want to be suspected to still share the old racist moral, is until today the foundation of the German nation state rationale that opened and opens the world for the post world war global German imperial policy agenda, an agenda which just coincidentally results these very days in the renewed offensive confession to the imperial mission of Germany obliged to nothing less but to intervene in the world of wars of the good against the evil.

Idealizing the mission of the German nation state after World War II as a new moral imperative and accusing the old one of being a moral failure, compared to what the very mission of the nation state supposedly is, is an idea that ennobled this nation state via accusing it failing to match with its genuine mission. The success story of this very typical social science critique not only proves its—

success. It is this logic, off-thinking the world of nation states even within wars and off-thinking the many ways nation states use and sacrifice their people for the global affairs about political and economic power and to critique the nation state, not for the prices people pay for this very genuine nation state rationale, but for their failed—moral—mission, failed missions social sciences even find in wars, it is this critique that preserves the trust in the mission of nation states that they genuinely must aim at caring about their citizens

The universalization of social science thinking......

With the post-World War II globalization of the US model of nation state and the transformation of the colonized world into nation states, making the whole world a world of nation states, social science theorizing was confronted with a paradox, a historical anachronism. Social sciences thinking that is thinking about and through the view of nation state constructs on the social, knowledge created from theorizing about the particular nation state socials in the imperial nation states was the knowledge about the world' social, across also those nation state socials of the former colonies.

Theorizing through space, the de-politicized phrase of nation state socials, is the epistemological notion and the topic of a discourse, under which social science theorizing, theorizing about nation state socials through the view of the nation state constructs, was completed as the global way of theorizing including social thought in the new nation states.

It was this global completion of social science as social thought across the universe of nation states that abolished the anachronism of nation state knowledge that represented the nationally constructed knowledge from the imperial nations as knowledge across the whole world's nation state socials—and it did this universalization of nationally constructed social thought with and thanks to the nationally constructed social sciences theories in the imperial world that rules the world's social thought, from there on also social thought in the new nation states.

That of all things, the application of the nation state society model to the colonies, the very model constructing a society that was responsible for making the colonies what they were, a means of the imperial nation states, would make them political entities on the globe, "independent" from the reign of the imperial nation states, is one of the odd tragic ironies, world history orchestrates. One must have witnessed how this tragedy was made a reality after world war II and the new reign of the US concept of a post-colonial globe of nation states, to believe that perceiving the creation of the new nation states from the former colonies was deliberating them and their people from being the material of the imperial world, was not meant to be bad historical joke. The wars, gaining this kind "independency", that made them more dependent than before, are telling everything, what this project creating independent states was about: Making the US Empire the ruling empire about a world of nation states constructed along their model of nation states.[20] Consequently, this historic tragedy making the colonies nation states, was repeated in the world of social thought: Once the new states were founded, the new nation states also applied the concept of social sciences to social thought, theorizing about nation states through the idealized rationales of nation states, a view that already guided the very illusion of a nation state serving its people, articulated in the paradoxical appendix of an "independent" nation state, including the institutional settings of Higher Education—and detected a historical anachronism.

As anywhere else in the history of nation states, it was the opposition against the imperial nation state and their social sciences that helped to abolish the anachronism of thinking in nation state views and of making the view of the imperial nation states the view of the whole of nation states and, by abolishing this anachronism to thus finalize the global reign of the system of social sciences of the imperial countries and to make this very system of social sciences the global system of social thought.

---

[20] Unlike the concept of nation state before World War II, the main new feature of nation state was, that the nation states were constructed to compete among each other about gaining their power means by benefitting from serving the global capital. The clearest measure putting this concept into practice is and was, that nation states had and have to finance the economic means for their political power via lending money from the global financial market they supervise. Subordinating their national currencies under the judgement of the global financial capital, was, oddly enough, interpreted as a loss of sovereignty, just as if it was not these nation states that set these rules for the world's nation states and for the global capital as the new global rules for their competition about political and economic power.

Not coincidentally, it needs scholars from former colonies, educated in social sciences theorizing in the imperial world, to articulate this historical scientific opera, advocating the need, to complete global social thought as a multiplicity of nationally constructed social science thought, abolishing the paradox of the new nation states, scientifically participating in a world of a multiplicity of nationally constructed knowledges with knowledge about the imperial national socials. Epistemologically, completing social science thought as a multiplicity of nationally constructed knowledges, implied to replace the claim for truth of theories in the social sciences by a multiplicity of contextual, of the many spatiological truths, the pluralism of relative local knowledges. [21]

Hence, the completion of global social thought under social sciences was completed by the erosion of what constitutes them as scientific knowledge, the erosion of their concept of objective knowledge, replaced by a global relativism of spatiologically constructed theories.

Accusing social science to ignore that theorizing is a matter of space is an odd misunderstanding about the social sciences and can only be articulated by very social science thinkers who learned their lesson to think from the social sciences that thinking is thinking about and through national constructs. To misread the universal claim of the validity of scientific knowledge of the "European" knowledge as knowledge that is lacking a spatial dimension, is the pretentious intervention, that aims at claiming under the epistemological notion that knowledge must be dependent on the space in which it is created, the creation of the very nation state way of theorizing and theories in the new "spaces" of the new nation states.

By opposing the notion of "universal" knowledge the opposition against the European science under the notion of "eurocentrism" and alike, social scientists advocating a multiplicity of spatiologically constructed knowledge, a debate also led by thinkers from the very "European" social sciences, opposed the last rationale element of social science theorizing, their insistence on a form of objectivity, and finally abolished what makes social science a form of scientific, objective knowledge.

Just as if there was any knowledge that was not local and universal, simply meaning that any knowledge is about somewhere

---

[21] See chapter C 176 ff in this book

and that this knowledge is always everywhere knowledge, the universalization of theorizing about national socials through the view of "spaces", advocated, to make any knowledge spatiologically constructed knowledge, that is to "provincialize" the European sciences as the sciences in the former colonies and, thus, completed a world of social thought consisting of a multiplicity of local relativisms. Conceding theories created in Europe which presented and still present the violence and exploitation of the colonies, founding the economic bases of capitalism in Europe, as a response to the nature of the "natives", as only one possible view, the "Eurocentric" view, is only possible by accepting the very European social sciences way of theorizing, that insists on the relativism of spatiologically confined social thought and does only allow a critique of any theory, including the "Eurocentric" theories, that relativates instead both critical thought as the critiqued thought and advocates exactly what the notion of "Eurocentrism" means, downplaying the—never as false critiqued—European theories towards only one spatiological, "provincialized", possible view on the social, that is opposed, not as false theories, but because it does not relativate its theories against other spatiologically confined and contextualized thought.

As a consequence, joining a world of a multiplicity of spatial relativisms in the first place detects scientificy, the objectivity of knowledge, as its main obstacle, and in the second place therefore opposes and abolishes scientificy, not ever critiquing any—from there on called—"Western" theories, for the sake of breaching their monopole on theorizing, thus paving the ways to also contribute social science thought to the world of a multiplicity of spatially constructed knowledges and, finally ends up where nationally constructed thinking about the world's social, consequently, ends up, in advocating global social thought as imperial thinking and its (sub)-alternative variations, preferably presenting the nation states of the former colonies, executing the imperial agendas, as the victims of a "Western" imperialism, they never want to find in their new nation states and, thus, finally establish theorizing about the world's social as the creation of a multiplicity of "spatiologically", i.e. idealized nation state views on the world's social, nationally constructed social thought, all claiming to be social thought that does not claim to explain the world but social thought that claims to rule spatiologically relativated theorizing across the world.

....completing the globalization of social science theorizing as a multiplicity of scientific patriotisms

Global social thought under the regime of social science in fact finally shifts the concept of thinking about and through idealized nation state constructs and the agglomeration of nationally constructed knowledge bodies as theorizing about the world's social one step further and social science thought from constructing it through national perspectives towards the concept of spatiological social thought, introducing the space, the "where" knowledge is produced as its cognitive resource and thus the knowledge this way of theorizing creates as the exclusive way of theorizing about national socials, scientific knowledge as the exclusive, cognitive habitat in a world of a multiplicity of parochial knowledges.

Heavily supported by some philosophical gurus from the imperial world, it was the privilege of the critical followers of social science thinking, especially in the developing world, or, to put it more precisely, social science thinkers claiming to argue in the name of the social sciences in the developing countries, to further develop social thought towards a politically motivated parochial opinion, finally thus burying the mere achievements social science thinking had gained from its—false—critique of classical philosophies. Arguing against their mysticism, arguing against teleological thinking that not only develops social thought from ideas, but interprets the social reality as substantiations of these ideas, spatiological thinking advocates thinking as articulating a patriotic voice about the wheres of the wheres, the home states, of which they consider their knowledge as voicing these view of the wheres as their contribution to global social sciences now consisting of the paradox of a multiplicity of *scientific patriotisms*.

In the modernized version of global social science thinking the mysticism of teleological thinking, social sciences in fact had overcome, re-appears as spatiological theorizing, the climax version of globalized social science thinking, in which politically constructed spaces, the many unique locals, are considered as the "context" through which the global social sciences voice a multiplicity of parochial thought, born by an exclusive mystic inclination of a local thinker with his local object of thinking, only an obscure inclination of a thinker with the "where" can think and articulate, a patriotism voicing the view of a member of a nation state that considers this membership as his essential social nature and his scientific mission..

Stating that any knowledge is created by a thinker some*where* and that any object of thinking is some*where* would be banal. However, in the modern globalized version of the social sciences the *where* of thinking is not at all banal, but it is considered as the object, the resource and the driving force of social thought.

> "... the proposition that thought is related to places is central to my project provincializing Europe"[22]

In the modern, postcolonial globalized social science approach to social thought *where* one thinks decides about *what* one thinks and therefore global thought under the regime of social sciences knows above all a distinction between spaces of knowledge, namely the one between inter-national and national and many other dichotomic couples of knowledge spaces.

To phrase it in different words: If the *where* of theorizing was not *the contemporary epistemological* concern in global social science theorizing, it would be a childish banality to mention this: There are phenomena which are here but not there. Any thinker from here or there reflects about the phenomenon, presents the theory the thinker created and shares this with other thinkers from here and there. In postcolonial social science thinking, this banality, the where, is not just the spatial aspect of the object of any thinking as it is time, but is considered as a cognitive force creating thought, a cognitive force that affects the contents of thought, and, thus, decides what we think about any phenomenon. Space, the where of things and the where of the thinker, usually politically constructed spaces of the social, more simply, nation states, have been transformed into cognitive actors through which social thought voices the thought of places, the thought of nation states socials.

In contemporary global social thought under the approach of postcolonial social science thinking the "where" of the objects of

---

[22] Chakrabarty, D., (2000) *Postcolonial Thought and Historical Difference*, Princeton University Press, p xiii

thinking and the where of the thinking subject craft thought,[23] and constitute the uniqueness of spatially constructed objects of thinking, a theoretical perspective, and a spatially constructed way of thinking, through which a spatially distinguished multiplicity of unique thought about spatially unique socials is created.

> "To 'provincialize' Europe was precisely to find out how and in what sense European ideas that were universal were also, at one and the same time, drawn from very particular intellectual and historical traditions that could not claim any universal validity. It was to ask the question about how thought was related to place. Can thought transcend places of their origin? Or do places leave their imprint on thought in such ways as to call into question the idea of purely abstract categories?"[24]

Discussing the most abstract category such as *"to ask the question about how thought was related to place"*, a question that, though it—seemingly—may not claim any universal validity, one might feel invited to ask the question, to which place, raising such an abstract question, is "related" to, a question aiming at questioning if thought can be thought beyond the "imprint" of places, if *"places leave their imprint on thought", a* question that obviously is no question, but a determined epistemological position, and this position is: Yes, they do. Raising questions that start from an answer will certainly find answers, confirming the answer such questions seek to answer. Which place should one ask this pseudo question, to be told by the place that it is place that "imprints" us their answers on our questions?

It must be an irony of social science thinking, that people working throughout their life torturing their mind to create thought about the social reality, are obsessed by the idea, that social thought must be an impact of the reality on humans mind, "transcending" their thought to the thinkers. Why then thinking? Since thinking "transcends" nothing but non-knowing by gaining insights the reality does not disclose or even voice to the thinker,

---

[23] While it is *the* cognitive challenge in scientific thinking to eliminate factors that mislead thinking towards any given biased thought, the postcolonial social science approach to thinking considers any thinking as the necessarily "biased" creation of thought and does though not terminate thinking that is not able to create any objective thought, but insists on the paradox of an objectivity of thought by disclosing the cognitive factors creating the biases of theorizing.

[24] D. Chakrabarty, (2000) Postcolonial... p xiii

even the "where" must be found out by thinking, it might be wise not trust that the mystic cognitive forces of the place, transcending thought about the relation between place and thought to this postcolonial thinker, but to trust the cognitive forces of human mind and to think about what this thinkers has heard from a "where" about how this where is related to thinking.

So, what are the kind of insights places disclose to thinkers advocating place as a source of thought?

> "Until I arrived in Australia, I had never seriously entertained the implications of the fact that an abstract and universal idea characteristic of political modernity everywhere—the idea of equality, say, or of democracy or even of the dignity of human being—could look utterly different in different historical contexts. Australia, like India, is a thriving electoral democracy, but Election Day there does not have anything of the atmosphere of festivity that I was used to it in India."[25]

It is always the same one and only argumentative game social science thinking is playing, ever keen to present their theories about the world as a theory voiced by the reality, here named place. The place is saying nothing—and the *observations* about let's say, as in this case, about "democracy", are observations the social science thinker *makes* about democracy in India and Australia.

Strictly speaking, stating that *"Election Day there does not have anything of the atmosphere of festivity that I was used to it in India"*, does not even say anything what a scientific thinker thinks about the difference between the enthusiasm here or the routine practicing elections there, he only describes as how people see elections. Distinguishing between the ways people in India and Australia *see and practice* elections in the first place wants to post the message, that without any argument neither about elections and nor about the relations electing people have to elections, that what elections *are* must be different between India and Australia, just because electing people *see* them differently.

Hence, so the conclusion of the social science thinker from what only he notices about elections, theories about elections in India *and* Australia, like any *"abstract and universal ideas"* must fail, because the electing people practice elections with a different view on elections. If this was the case that is how people see elections decides about what elections are, how then does Chakrabarty know,

---

[25] Chakrabarty...,p xii

that both things are though the same abstract thing, elections and not something entirely different? Social science only play with pointing on the reality as a point of reference to prove what their *theories* about the reality are, to make their theories about the reality they see, taken as being voiced by the reality, and thus, undisputable. It is the reality they cite as a witness for the objectivity of what only they think about the objects of thought, the prove for the undisputable knowledge voiced via the thinker by the reality, ironically while proving the necessity of a relativism of thought dependent on the space where thought are created. What is indisputably proved for Chakrabarty, is that place disproves the theory of *"an abstract and universal idea characteristic of political modernity everywhere", of* an abstract idea, here elections, an abstract idea he though argues about by saying that they, the abstract ideas, *"look utterly different in different historical contexts".* Yes, they may look different, but an observation of a difference in the way a something is looking *argues* with this very abstract something that is considered as being essentially the same and that has different ways of practicing this abstract same thing, otherwise he could not even compare them as variations of the same abstract category election.

However, what Chakrabarty and all the social science thinkers, advocating the cognitive power of place, want to say is that it is a theoretical mistake to create any abstracts ideas that tell us something about the same thing in different places, may this be with different interpretation of the same. They argue against what abstract categories mean as "tabula rasa", discrediting their abstractions they know very well while arguing with them as meaningless, with the false argument, that a category is not the same as observed things. With their observation that elections are seen and practiced in different ways in India and Australia, they want to say that the category "election" is meaningless, because it abstracts from what makes a category a category, from their differences of the same— category.

A cat, is a category, and these social scientists want to persuade us, that due to the fact that there are white and black cats, they very well know to distinguish from dogs, that the category "cat", they very well use for their distinction into two kind of cats, is meaningless, because they falsely interpret the abstraction any category makes as extinguishing what both animals essentially share, because they do not share the same color, just as if they want to say that, since anything white was a cat and anything black was a cat, cat cannot be category. In other words, social sciences argue to backdrop scientific thinking behind the preconditions of thinking already any language and its categories achieved and provides for thinking and, thus, to oppose thinking and to advocate their category of a place "imprinting" knowledge to the thinker, in order to advocate their theory that thinking must be articulating patriotic views about voicing the view of their home land.

The advocates of the place as a cognitive instance argue against what constitutes scientific thinking, if not already the abstractions already languages make, creating abstract judgments, saying something about what different things, may this be in different places, share, such as about elections, and advocate that theories must vary dependent on the place where they are produced, thus, opposing what scientific knowledge essentially is, that knows to distinguish between, let's say, what elections are and how they are perceived in different locations. It is what makes thinking scientific thinking and even already language does, abstractions from the differences *known* as variations of the same, such as elections in India and Australia, the advocators of the place as a cognitive instance want to oppose and overcome, because it is this essential of scientific thinking, the categories materializing knowledge for thinking, they detect as an obstacle for theorizing, that aims at theories representing the particularity of a place.

One may very well argue about the abstractions categories make, however, the theory about place arguing about place as an epistemological dimension argues against categories as such to oppose what constitutes theorizing, not only scientific theorizing. It is the scientificy that disturbs the project advocating to create theories, which contribute nationally constructed knowledge to the global social science thought representing a particular nation state view, here a view about and through India, they present—oddly enough—, not as the political desire for an authentic nationalism, but—since they are no politicians, but scientists—boldly, as an epistemological must of scientific theorizing, and present, by the way, a very "*abstract and universal idea*", about place as a cognitive instance in theorizing, though a very false idea, here about social science theorizing.

Needless to say, as a scientists, that is a thinker creating categories by abstracting from non-essential differences, these thinkers about place arguing about place as an epistemological instance of thinking, aiming at relativating any thought as dependent on place, insist that theorizing as voicing the view of the "where" is not only a plea for thinking about India, and not only any "local" category they were voiced by their very demand for advocating local patriotisms, but that patriotic knowledge is not only an spatially confined epistemological must for a particular place but a spaceless truth that counts independent from any place everywhere:

> "If this argument is true for India, then it is true of any other place as well, including, of course, Europe or, broadly, the West."[26]

Categories, once they articulate what postcolonial thinkers were "transcended" by one place, India, postcolonial thinkers know and enjoy what categories are, they just denied to say anything, that they are meaningless abstractions, "tabula rasa". However, if the category is saying that it is the nature of theorizing that it must always and everywhere voice patriotic views about nation states, then categories are not only meaningful, but the advocates of patriotism, passionate fighter against categories and even more against universal categories, detect essentials of science, objectivity, and are bold enough, to claim that interpreting science as voicing local patriotism is a universal truth about social science thinking. This, postcolonial social science theorizing about global social thought in the social sciences, advocating that global social science thinking must be a multiplicity of patriotic thinking across the whole world, cannot be accused of being a genuinely imperialist concept of global social thought advocating competing very national views, because it is only the epistemological necessity of the mere geographical category space voiced by space.

---

[26] Chakrabarty,... xiii

From Marx to Heidegger:
Critical theorizing in the anti-colonial movements—self-purified for constructive imperial nation state views

It is another irony, accompanying social thought that advocates the need for nationally constructed theory bodies in the former colonies, presented as the epistemological necessity of the locus where social thought is created, though nonetheless quite consequently, in order to support the argument for a mystic cognitive power of the "where", another name for nation states, here India, to argue about "provincializing Europe", to support this theory with the most European of all European thinkers, with the German "*Blut und Boden*" philosopher Heidegger. If one wants to justify that one's national patriotism needs a scientific voice in the global concert of nationally constructed social science theories, and if a thinker has the temerity to present his patriotism as the natural effect, as a necessity of place, only odd Marxists and other thinkers, insisting on science as "abstract and universal categories", refuse to accept, than Heidegger is, no doubt, the best philosophical advocate for theorizing as celebrating ones attachment to a home country.

It is a very particular double irony in the history of the world's social sciences that it is the correct opposition of academics from—or at least arguing in the names of the developing countries—against a false critical theory, it is their critique of the "Historical Materialism", an ideology the Soviet Union had established against the Western ideology production and spread across the world as an opposing world view, mainly into the colonized world supporting their fights for independence—that it is this correct *critique* of the Soviet Union ideologies created in the European social science in the postcolonial discourses, created in the name of the former colonized world that has helped to established a false theory about *space, the* epistemological creature of the European social sciences and their inclination to think about any social through nation state perspectives, to universalize this false theory about space as an epistemological must finally as a worldwide acknowledged epistemological dimension of theorizing about the world's social, more than the universalization of the European social science concept of theorizing itself ever could during their colonial reign.

Chakrabaty's arguments stands for an essential theoretical error of an opposition against the "Historical Materialism"—a historical epistemological twofold accident, the wrong self-critique of a wrong theory, the ideology of the Soviet Union, an interpretation of Marx as a type of scientific religion—that critiques Marx via— though only—rightly—critiquing a caricature like Marx interpretation, that could not more contradict Marx's theories. It was already Engels, who was often critiqued by Marx, when Engels used the term "Marxism," for Marx's theory about capitalism, thereby interpreting Marx's theory about capitalism as an *approach* to science, who perpetrated the caricature like notion of the "Historical Materialism."[27]

First, uncritically adopted by social movements against colonialism and used for the independence of colonial countries, then, once the former colonies became independent, now critiqued by the same academics in the new decolonized states as an intervention into their *local* national identity-building process, this science ideology of the Soviet Union was then seen by theorists from the "periphery" as a theory that did not know any "where," that is, a way of theorizing that did not allow the creation of their own *local/national* worldviews. And ironically, this, the critique of the "Historical Materialism", is a correct critique, a critique that only occurs to foil a religious like theory, counteracting and trapped by the dogma of the *Historical Materialism*: In fact, the Soviet ideology subsumed the whole world under its odd dogma of an *historical*

---

[27] Deterministic thinking, opposed by Chakrabarty and at the same time practiced and advocated with the notion of space as a resource of thought, is, of course, not the privilege of the Historical Materialism. The social sciences are rich of examples presenting humans will and actions as the impact of the social reality, the whole idea of Behaviourism constructs will and humans' actions as a reaction on being. However, advocating a pluralism of relative knowledges that allows the former colonies to create their own nationally driven social sciences theories, sounds more convincing while advocating the implementation of the very model of nation state, that was responsible for what the colonies were, since it needs to transform the opposition against the colonisers into congenial partners as a new member in the global family of nation state thinkers—especially if one aims at getting the sympathies for the project building a nation state from the new global supervisor over the world of nation states helping to get rid of the colonizers in the arena of theorizing.

automatism, in which the authorities in the Soviet Union did themselves not believe, the fighters for independence opposed, since this dogma made them to supernumeraries of history and imprisoned their thought into repeating believes they could not identify with what happened in the colonized world.

In fact, if they had ever read Marx, they would have noticed, that *the essential elements* of capitalism were spread across the world as the world's ruling economic system, subordinating the world under the aims of this economic system, but that this does not at all mean, that this system introduced the essential elements of this system into all countries across the world. The economic system of the colonized world did and does not need to work as the same economic system for the world's business as long as it serves this business, by obeying some of its fundamental rules, may it just be by selling their goods on a global market. And if this selling towards the imperial world violates the rules of capitalism in the imperial countries, because in the case of the former colonized nation states the market economies are rather an odd version of market economies, in which the buyers and the sellers are the same, both private property owners from the imperial world and the nation state is rather legalizing such political-economic relations violating even the rules of the market economy, this very much fulfills the economic interests of the imperial world. The imperial capitalism is no fanatic ideology that insists on a global validity of its rules but a way to achieve its global material aims, and where achieving these aims violates their own rules, as long as it provides the business with profit, these rules may count when and where they serve this aim.

Hence, confronted with the dogma of the Historical Materialism, stating that the world follows the same economic laws, theories in the developing world rightly opposed this dogma and did it, though, with a wrong conclusion. Insisting on their local economic phenomena, instead of critiquing the theory that falsely subsumed the in fact existing differences between practicing capitalism in the imperial and in the colonies and the new nation states under their automatisms of any global laws ruling the world, they rightly insisted on creating theories taking the local phenomena into account and from there opposed even correct theories about global capitalism to develop their pretentious critique towards what they were after all aiming at. As one can see from those arguments like the ones from Chakrabarty about "place", this critique was and is not about critiquing false theories and their false abstractions about the imperial capitalism, but a critique that is guided by the idea to develop their dogma of spatialogical thinking, and thereby to justify the need for the creation of patriotic theories, justified as an epistemological necessity, a theory about thinking that ironically repeats the very deterministic logic of the Historical Materialism in the field of epistemological reflections.

Spatiological thinkers are good apprentices of what they believe to oppose. In fact, the idea, that it is the place of theorizing that determines thought, is a repetition in the field of epistemologies, no other theory but the very "Historical Materialism" stated about the history of mankind. However, repeating the very same logic, that make human's will and actions a necessary impact of the reality, this is the logic of both. The fact, that their critique of the Historical Materialism rightly detected in the false logic of this way of theorizing the very logic they apply to their theory about place inducing knowledge, does, however, not bother this critique, as long as it helps to justify the need for a particular patriotic nation state view in the new nation states.

The theory introducing the "where" as an epistemological dimension of social thought since then has flourished worldwide and crafted the main stream opposition against what was from there on called the „Western" sciences, social sciences since then coined with the space they origin from, opposed under another spatiologically constructed notion, "Euro-centrism", and inspired the creation of patriotic theory bodies under the notions of "local", "indigenous" or "Southern" theories and the like.

It is obvious, that social thought consisting of this kind of spatiologically constructed knowledge about their knowledge islands is never about knowledge, but about a purposefully constructed politically "biased" patriotism and it is this type of relativated knowledge the universalized social science world has learned from their scientific colonizers, both from the imperial nation states sciences as from the Historical Materialism. Both successfully universalized their dogmas of thought as an impact of the reality, the Historical Materialism with his determinism, the European social sciences with their proofs of knowledge against the empirical reality, variations of the essentially same dogma, and the postcolonial thinkers, arguing in the name of the former colonized world, applied the shared essential of these dogmas to their theory about the where of the reality as an epistemological dimension of social thought, advocating a multiplicity of spatially preoccupied theories, a multiplicity of spatially relativated social thought that gives the new nation state thinkers their space for their provincialized theories about the world's social. Hence, they critique theories they oppose under the notion of "Eurocentrism" as one sided views, they therefor propose to "provincialize", allowing them to create their own provincial theories as a contribution to global social thought, consisting of spatiologiallly pre-occupied world views, and consequently critique the "Western" theories never other but claiming to complement these opposed theories, all downplayed towards only one "provincial" view on the social with their own spatially relativated theories. Critiquing these provincialized theories as false thought is the least option that would come to the minds of provincialized thinking. They know from the colonizers that theories must be a matter of any theory model one choses for theorizing and that theories therefore must always create relative knowledge, thus, they argue, that since theories are the result of any ex ante choices for theorizing, they complement the methodological relativism of the European social sciences and the determinism of the Historical Materialism with the relativism of the synergy of both, and argue for space as the driving force of theorizing creating spatially relativated and determined knowledge, advocating a social science world that must commit itself to create locally pre-occupied and relativated social thought.

Hence, under the auspices of this postcolonial spatiological thinking, indigenous knowledge[28], once the discriminating notion from European sciences, excluding this knowledge from sciences via defining what scientific knowledge is beyond any argument about what indigenous knowledge said, is now re-discovered as a

---

[28] Indigenous knowledge is not a creature of indigenous people, but of the scientific colonisers. The critique indigenous knowledge was autochtonous is a revealing hypocrisy of thinkers who otherwise cultivate contextualized theorizing. The fact that all those indigenisation enterprises are the enterprises of the incriminated academics which serve the latest epistemological fashions of the apparently not only Western "poststructuralists" relativists freaks, does not bother the original holders of this indigenous knowledge: For the disappointment of the indigenous knowledge seekers they don't care about what indigenousness at all means.

form of spatiological knowledge, that also does also not want to critique any theories it opposes, but aims at contributing another local unique knowledge island with an exclusive local theory: For this very purpose, creating scientific patriotisms proving that the colonized world or otherwise from the European social sciences excluded theories as in the case of China, that they already owned spatiologically constructed theories long time before the "Western" sciences invaded the world with their "Eurocentric" theories, some social scientists interpret thinkers, preferably social scientists opposing the "Western" social sciences, detect even pre-social science thinkers like Khaldoun, a thinker, thinking about the 14th century, as the first spatiological sociologist; others admire poets like Rizal, insisting that Philippine people are not lazy, as a local anti-hegemonic theorists, all re-detecting the colonized world as a world that provides *authentic local knowledge,* all now labelled as "Southern" theories for today's global battle about the locally preoccupied theories, all representing the authenticity of a national social view in the battle about patriotic theories, which do no longer claim to understand the world, but to represent an exclusive national view, a scientifically ennobled local patriotism as their contributions to a world of provincialized theories, not aiming at understanding the world' social but arguing about the scientific prerogative interpretations of the world social with their local patriotisms.

It was and is this opposition against the theories—called in the logic and language of spatiological theorizing—"Western" theories, that paved the way for universalizing the "where," presenting space, mostly politically defined, as a worldwide epistemological dimension of science, thus extending the concept of social science thinking about and through nation state view across the world in a way, the sciences from the imperial world never could, due to their own very political, i.e., nationalized "where" of theorizing, that basically did not care about what is going beyond their biotopes.

Since then, thanks to a false opposition against the imposed "Eurocentric" theories and justified by the very European Gurus of post-structuralism, social science thinking finally conquered the world of social thought as a multiplicity of relatively obscure spatiological social thought, only comprehensible for and through the obscureness of being part of a "space" in which it is created. Since then, the world of social sciences consists of the global theoretical relative crux of anywhere valid and nowhere sharable theories and an intellectual scenario, in which the traditional obscurantism of religious thinking, once overcome by social science thinking, rises again from what the European social sciences had passed on to the colonial science world.

Thus, it is the false opposition against the universalization of the theories from the local knowledge island in Europe, which further develops global social thought under the regime of social sciences, and which proceeds to oppose the essentials of scientificy of social science thinking for the sake of setting free the creation of all the local obscurantisms and, thus help to abolish the only substantial achievement thinking about the social in social science thinking has gained through its critique of the classical philosophies—with the help of a consequent interpretation and application of the dogmas of social sciences thinking to global social thought.

Global social thought in social science thinking returns to the obscurantism of thinking in which the mystification of some spatial particularism into a type of knowledge can be only shared by those who share the mystic spatial "context", in which it is created. It is this scientifically reactionary opposition against social science thinking that gives birth to the emergence of a new wave of the paradox of religious social sciences some centuries after the European sciences emerged from their critique of the obscurantism of religious thinking in the medieval times.

...opposing a monopoly on spatiological thought in the "centres"

Once global social thought became a multiplicity of spatially constructed theories, presenting global social thought constructed through national perspectives is not rejected in social science thinking as an ordinary theoretical fault, as explicitly preoccupied theorizing, theories that violate the scientific criteria of the very social sciences, global social science thought admittedly represents patriotic ideas perceived as scientific theories; from there on global social thought under the regime of social science considers boldly nationally preoccupied thinking in global social science thought even as a reflexive sine qua non, without which social science thinking would not be able to understand the particularities of the state biotopes socials and which are since then considered as the only knowledge contribution to global social science thought. Theories which might claim to be constructed as coinciding with what social sciences consider as scientific thought, as knowledge that is not relativated due to its spatial context, must accept that they are also only the same kind of preoccupied patriotic thought and that claiming them as science is only claiming a monopoly on theories among the pluralism of relative knowledges. Constructing theories not only through a national perspective, but creating purposefully theories that represent a multiplicity of "local" particularisms, is not only not considered as a the only possible way of thinking, thought imprinted by place, creating purposefully nationally preoccupied theories for the inter-national knowledge arena, is the objective of global thinking in the social sciences all constructed along the postcolonial concept of spatiological knowledge.

Consequently denying any shared systemic essentials of the different nation states rationales, in the first place their shared concept of all being nation states, their insistence on the national local peculiarities as the nature of knowledge about the secluded national socials implies that global social science thinking introduces thinking through the national peculiarities as an epistemological dimension of theorizing about the individual national social, off-thinking what they essentially share.

> "No concrete example of an abstract can claim to be an embodiment of the abstract alone. No country, thus, is a model to another country, though the discussion of modernity that thinks in terms of "catching up precisely posits such model."[29]

It is true that social sciences thinking thanks to their national perspective through which they theorize about national socials, consider their theories as true theories. And it is this what distinguishes scientific knowledge from other formats of knowledge, which do not claim to be true and do not need to prove their thought. Discriminating the fact that these theories claim to be scientific theories and therefore, in fact, to create knowledge that must be proved as true knowledge and not just as any arbitrary opinion, and to identify the *scientific* claim for truth with the *political* claim as being "models" to another countries" social, is howev-

---

[29] Chakrabarty, ...xii

er a pretentious misinterpretation of scientific thought and truth, however, a misinterpretation the particular way the social sciences theorize, may insinuate, a misinterpretation, however, only for those for whom theoretical abstractions are the same as imperial politics and for whom scientific theories are the same as the articulation of political statements.

The fact that social sciences theorizing creates theories about the social reality ever starting from assumptions, they mostly call "hypothesis", and which they present as theories proving their thought with certain cognitive procedures as true knowledge, requires to disprove these theories as false theories, false theories because they do not construct their theories from the reality, but from any ex ante presuppositions through which they see the reality. Proving or disproving theories originating from assumptions therefore needs to trace their thought and show if and how they are not true theories, but theories, which thoughts are only the result of the pre-assumptions through which they think about the reality and, thus are disproved as false theories, proved along their thought. Identifying the claim for truth with the claim for a "model" of living is a pretentious misinterpretation of what science is about, this mistake can is a very purposeful constructed identification in order to oppose any imposed model, because this opposition against truth is the opposition against models of living, because it wants to insist on other models, other politically constructed models of any social, argued for beyond any needs for scientific argument against any theories and there for opposes scientific arguing as if it was the same as opposing models for living.

Accusing "Western" theories to impose models for theorizing, not opposes this way of preoccupied thinking resulting in false theories, but it opposes arguing scientifically in order to advocate liberating scientific thinking from its scientific procedures to open scientific thought for any preoccupied thought beyond the scientific rules of social science theorizing, open for any patriotic view, and therefore suggests to abolish even the scientific format of social sciences theorizing, aiming, not at critiquing any of those preoccupied theories, but at opening social science for thinking through any "model" any scientist may sympathize with.

In this sense, postcolonial thinkers are in a certain sense honest scientific liars: Let' s forget about pretending that our thinking through ex ante models contains any objectivity, let's admit that they are anyway only pre-occupied thought and therefore lets free scientific thinking from what only prevents a multiplicity of national "models" to guide the constructions of theories, representing a multiplicity of patriotic theories, without getting stuck in proving their truth and, consequently, thus have a multiplicity of truth dependent on the many wheres, the many nation states theories. To open social sciences for patriotic thought as scientific thought social sciences must eliminate what constitutes scientific thought, a traceable prove of the objectivity thought.

If scientific social thought is considered as dependent on the location, then the fundamental idea founding social science thinking that scientific social thought must prove its objectivity is dissolved into the multiplicity of locally dependent truths, and any social thought can be presented as true scientific thought. It is true, social sciences do operate with a society model they consider as the nature of "civilization". However, advocating a way of theorizing that dissolves scientific thinking into the arbitrariness of politically biased theories is neither a critique of the "Western" social science theories, nor is it any critique of the society model on which these theories are founded. Advocating patriotic thinking is rather the opposite: Not only is the theory about thinking voicing the place a false and obviously pretentious theory about social thought, not only does the plea for spatiological thinking abolish the only essential achievement social science thinking at least in a formal sense, its insistence on objective social thought, how false their concept of objectivity may ever be introduced into social thought, not only does spatiological thinking abolish exactly this achievement of social science theorizing, it also does not at all provide the slightest critique of the society model the former colonies instead imitated once they gained their independency and became the very same nation state societies as the incriminated "West". The only thing spatiological thinking achieved is the scientific entitlement to create patriotic views on the social world as scientific knowledge—for the price of deteriorating this very scientific knowledge towards a pretentiously politically biased, relativated opinion.

Thanks to this intervention of spatiological thinking, thinking through any national/local "perspective" is no longer considered as violating the nature of scientific theorizing, but is from there on in the global social sciences considered as the an acknowledged way to grasp the nature of the social of the individual state social, representing national particularisms as the mission of theorizing for the encounter of knowledge in the inter-national theory arena.

"Eurocentrism", *the* contemporary critical notion against the claim of "universal" theories, accusing them of monopolizing theorizing about the world with the body of the European social science theories, thus, not critiques the idea of theorizing about state social through national perspectives as thinking that enthrones national prejudices as the analytical framework for thinking about the nation states social, but advocates the creation of purposefully "biased", politically inspired social thought as the mission of scientific thinking about the world of nation states.

No doubt, it is the European sciences, which have imposed their views about the world on the world. However, making the opposition to the European sciences a matter of a competition about the global scientific leadership among a pluralism of patriotic theories and shifting the theoretical *leadership* from the European monopoly toward a multiplicity of parochial world interpretations, does not only leave the "Western" theories un-critiqued; just as if they want to ignore Said's warnings, this critique of elsewhere locally non-applicable theories rather aims at replacing the Western theoretical monopole on nation state theorizing by the multiplicity of spatially restricted and politically biased ways of theorizing, and is thus the final real globalization of the very concept of thinking in national social entities through national constructs, a way of thinking social science thinking introduced and imposed to the world. Thanks to spatiological theorizing now the European approach to social thought in the social sciences really becomes universal, no longer imposed to the world, but thanks to the most active support of those who might believe to critique the European social sciences voluntarily expanded to the whole world beyond the European social science traditions, they rather apply to their needs of new nation states and their demands for authentic national views and radicalize the relativism of the European social sciences by extinguishing the last formal elements of science from the social sciences, they consider as an obstacle to join the European approach to social sciences with their own nation state view on the social. It is another irony of the world social sciences that the notion of "Eurocentrism" standing for a main strand of criticism helps to apply and to radicalize and, thereby, to finally globalize the very approach to social thought in the social sciences, these criticists believe to oppose, though the only thing they oppose is the monopole the European social sciences have on the European approach to social thought.

The simple but most fundamental false conclusion founding this critique phrased in the notion of "Eurocentrism" is disclosed by this notion under which it is rightly summarized: It is an opposition against the theories coined as "Western" social thought that learned so much from the European social sciences that it opposes their monopole on theorizing about the world, an opposition which apparently decided that it is neither the content of the "Western" theories spread across the world from the "West" nor their approach to theorizing, they should oppose but instead, to oppose the monopole the theories created by the European social sciences with the European approach to social thought hold.[30] Only this opposition, aiming at questioning this monopole coins the location, the "Western" theories as a synonym for a critique, that argues against the location where theories come from and therefore the locally limited view it must represent, argues for gaining the acknowledgment from the very "Western" theories, that they are only theories from a single "where" and thus discloses that this critique aims at being a part of the very social science, applies the very same approach to social thought and complements the very kind of theories from "West" with another provincial body of social thought from a "global South".

And this, the opposition against the monopole on theorizing and the claim for alternative theories, acknowledged by the very "Western" theories as alternative views on the world, has a tragic

---

[30] The notion of "The Northerness of Globalisation Theory" is not a theoretical accident, but a major criticism of—"Southern Theories". "This body of writing, while insisting on the global scope of social processes and irreversible interplay of cultures, almost never cites non-metropolitan thinkers and never builds on social theory formulated outside the metropole." (Connell, R. *Southern Theories*, Polity Press 2007, Cambridge) That is true, this is "a body of profoundly limited" theories—however, complementing a "Northern" body of theories with a "Southern" body of theories is not only not the same as analysing "*the global scope of social processes*"; adding to a spatially confined body of theories an other spatially confined body of theories—is just adding two bodies of confined theories and does not overcome spatially confined thinking in any theory body, but adds another body of confined Southern theories. Confined plus confined is the same as analysing the global scope of social processes?

implication, tragic at least for those, who believe that their opposition is an opposition against "Western" theories about the world's social: This false critique is indeed succeeding to replace the monopole of the very "Western" theories, imposed via the colonial power across the world as the only way of theorizing, which now, by applying and developing the approach of nation state theorizing of the social sciences towards "provincialized" theories finally succeed to establish the "Western" theories as the global reference theories and the European approach to social thought as the reference approach for theorizing about the world's social in a post-colonial science world, the European sciences started to accept and coined very meaningful as a "multi-polar science world", a science world that consists of a multiplicity of locations of the same concepts of social thought—variations of the European social sciences.

.... liberating global social thought from scientificy for creating patriotic theories

Universalizing thinking about the world's social with a multiplicity of explicit nationally constructed categories and theories arriving at a multiplicity of patriotic social science theories, is not yet the complete story about global social thought under the globalized regime of social sciences, that abolished a monopole on spatial theorizing by a multiplicity of nationally constructed views in a world of nation state socials.

Armed with the postcolonial liberation of global social thought from what the social sciences otherwise critique as "biased" thinking, spatiological theorizing becomes equipped with epistemological insights some discriminated "Western" epistemological gurus created, supporting the idea of spatiological thinking saying that knowledge must ever be contextualized. It is the very "Western" social science gurus who argue to dissolve any distinction between the object of thinking and the thinking subject and who suggest that it is the object of thinking that must be seen as the cognitive agency that creates thought that are only voiced through the mind of the thinking subject and that, hence, the articulation of prejudices as the nature of global social science theorizing.

Not only "...*that all understanding inevitably involves some prejudice....*" is the insight "Western" thinkers like Gadamer feed into the debate about global social thought, this "Western" expert for international knowledge sees such prejudices not as false knowledge social science theorizing should critique as prejudices, but he considers *prejudices as "conditions of understanding"*[31].

The fashionable habit talking about "knowledge production" while theorizing about social thought just as if thinking knows its result in advance as it is the case in producing anything, in global social thought under the regime of social science the notion "knowledge production" becomes an appropriate image for the way of pretentiously theorizing in global social sciences, in which the creation of prejudices is considered as a *"condition of understanding"*.

Global social though in the social sciences, indeed, finally liberates social science theorizing from *the* epistemological paradox of social sciences theorizing: While social sciences not only accept but heavily argue for the epistemological necessity of a pluralism of the whole set of theories, they insist at the same time on proving the truth of any individual theory—except in global social thought. The parallel existence of a pluralism of relative and of true theories is what global social thought in the social sciences overcomes by extinguishing objectivity as an essential epistemological element of social thought and propagates the sheer relativism of spatiological theories, representing a multiplicity of patriotisms as the scientific body of global knowledge, prejudices produced for this very purpose and therefore with the explicit objective to produce theories, in which their messages are not only known before theorizing, but which also do not need to be proved. More precisely, since this scientific exercise to create admittedly biased knowledge insists to be scientific knowledge, the proof of this knowledge, unlike elsewhere in the social sciences, is its uniqueness, the in-repeatability of its thought beyond the where it is created.

Local, global, national and international, southern, northern, western knowledge; space in this variation of global social science thinking is not seen as any incommensurable feature of thinking, a contradictio in adjecto, such as a yellow distance, but considered as an new epistemological necessity in social science theorizing for the international science arena, constituting a cognitive perspective through politically motivated and culturally obscured pre-

---

[31] H.-G. Gadamer, (1979) *Truth and Method*, London, p 239

assumptions, through which the objects of thought must be approached, transforming space from an object of thinking into a theoretical "paradigm" for theorizing, in order to create spatially unique knowledge as a contribution to global social thought: Though all the debates about global, local and glocal show that space as a cognitive instance of thinking is an incommensurable attribute of thinking, the sheer determined will to create knowledge that represents the scientific absurdity of a national knowledge identity, constructs knowledge against all the epistemological features of social science knowledge as the theoretical paradox of spatiological knowledge for the global social science encounters.

No surprise that this scientific paradox does not work, or better, that this monstrous concept of scientific theorizing only works by arguing against a disturbing essential of scientific thinking:

> "In social sciences Zahra Al Zeera critically reviews the conventional positivist paradigms in the west. She finds that emergent paradigms of post-positivism, critical theory and constructivism have provided some space for alternative ways of thinking and understanding. She suggests that they are nevertheless connected by an 'invisible string' to Aristotelian principle of 'either/or', which holds that every proposition must be either true or false. This principle fails to integrate the material, intellectual and spiritual dimensions of life...This is where Chinese traditions of unity, harmony and oneness can play a role. Potentially Chinese efforts to indigenize its social research can make important contributions to a re-balancing of Western and Eastern patterns of knowledge ...In social research indigenization means to integrate one's reflections on the local culture and/ or society and/or history. ...Chinese researchers need to develop their unique perspectives and values based on their rich local experience, an awareness of their local society and culture.... Chinese social researchers need to respond to the momentous challenge, rather than taking the rationality and progressiveness of science as an obvious fact. "[32]

Spatialized theorizing detects in the first place the essential of science as its major obstacle to create thought that do not serve the need for knowledge, but for knowledge that represents the uniqueness of the a national society in global social thought and rightly discovers that the progress the epistemological departments of the discriminated "Western" social science indeed provide the justifi-

---

[32] Yang, Rui, Indigenised while Internationalised?, Tensions and Dilemmas in China's Modern Transformation of Social Sciences in an Age of Globalisation, in: Kuhn, M., Okamoto, K., *Spatial Social Thought, Local Knowledge in Global Science Encounters*, Stuttgart 2013, p 55/56

cation for the paradox of an anti-scientific knowledge serving the knowledge needs of internationalized national theorizing.

It does indeed: Supported by the epistemological affirmatism of the *"paradigms in the west"*, that thinking must be ever contextualized thinking—of course except this context free global dogma about contextualized thinking —spatial or indigenous thought compliments to the preoccupied thinking of social thought constructed through the view of nation states on their objects of thinking a variation of social sciences thinking that introduces intentionally preoccupied social thought, clandestine unique thought that can only be created and perceived by those who share their pretentious pre-assumptions. Spatial thought thus radicalizes to dissolving the world's social thought into pretentiously exclusive theories and thus radicalize the disaggregation of nation state knowledge islands with the disaggregation of local prejudiced theories, in which any *where* counts as a spatial particularism that claims this particularism as cognitive perspective through which it constructs its contextualized presuppositions, prejudiced, pretentiously obscure and exclusive theories created as a component to global social thought.

As if opposing views and theories, wherever created about whatever phenomenon encountered at which ever location, was not the point of departure for theorizing, for digging into what things really are, what they have in common and what not, the spatilization of social thought introduces the creation of a diversity of unique views about any social phenomenon not as a tribute paid to the imperfectness of humans mind, as the social science epistemic debates insisted, but declares nationally constructed prejudices as the final desirable aim of thinking and justifies to therefore never arrive at any shared knowledge as a necessity to match social thought with the uniqueness of the "place" it represents. In other words: According to spatial theorizing true knowledge that matches with the unique nature of the object of thinking must be therefore—the absurdity—of exclusive, never sharable, knowledge.

Spatial theorizing argues that phenomena must consist of any spatially unique identities, requiring the need for a spatial relativism of social thought, justified with the spatial peculiarities of the object of thinking, this dogma apparently not only knows as peculiarities distinguished from other spatial peculiarities, but even knows them before knowing the object of thinking as an approach for thinking about them.

Concluding from the mere fact, that there are certainly many things that only exist in a certain place, that knowledge and sharing this knowledge about them must be bound to a being in this location is the final end of a debate the social sciences began with the odd idea of a scientific universalism, an idea, only social science thinking can create.

Only an approach to social thought that decides if knowledge is knowledge or not, not depends on the coherence of its reasoning but depends on the extent to which it is shared by others, an approach that considers knowledge as a reflex of the object of thinking voiced through the thinkers, can create the idea that if knowledge is a matter of its spatial spread, to arrive at claiming that true knowledge is the same as universal knowledge—and consequently provokes the false critique, that essentially founds the false opposition against—consequently—spatially constructed theories: The opposition against the "Western" theories, the prevailing criticism in contemporary global social thought is the concept of local or indigenized theories.

Only this—false—critique of the concept of a scientific universalism that shares this equalization that true knowledge is universally shared knowledge, opposes the claim of universal knowledge, creates alternative knowledge and does not ever think about critiquing the faults of the theories it opposes, thus, after all, not disclosing why it opposes these theories, except that this knowledge must be knowledge that *represents* any other "where". Only an opposition against social thought that shares the view that knowledge depends on the extent in which it is shared can be trapped by the idea of universal knowledge and opposes the spatial claim of universal knowledge by advocating a multiplicity of spatially, that is nationally constructed and locally confined prejudices. Globalizing patriotic theories opposing a scientific universalism is the desolate motto of this opposition. Hence, this odd opposition opposes this knowledge *and* accepts it and compliments it with other knowledge of the same kind, insisting that both are spatially confined and contextually biased, because they are bound to space, no matter what this knowledge is about, no matter what the knowledge is saying and no matter where this knowledge is perceived.

To put in in other words: Considering that, with a few exceptions[33], todays social sciences across the world are so keen on being part of the very European social sciences, opposing their claim of being universal theories, is another odd historical irony—and a false theory or a pretentious misunderstanding about the notion of a scientific universalism. No doubt, the European social sciences, as the opposition against this notion phrases it, considered and still consider their way of theorizing and their theories *"as a unique and superior kind of knowledge"*[34]. Yes, it is surely the case, that the colonizers presented their knowledge *"as a unique and superior kind of knowledge"*, however, concluding from the fact that the colonial powers forced the colonized world to share the views of the colonizers, that it is the European science that imposes a scientific universalism is an odd identification of the political power of colonialists with an only imagined power no knowledge has and misinterprets this, the identification of the political power forcing the colonized to share the views of the colonizers, with the epistemological claims of what scientific knowledge claims to be. Yes, under the regime of the colonizers rejecting their theories without having the power for doing this, was not possible. No doubt, the colonized people were lacking the power means, the Chinese emperors had and used to simply ignore the attempts of the Christian missionaries to missionary China. However, only if one identifies the colonial power that insists on a superior kind of knowledge with a mere epistemological notion that says that any knowledge is universal knowledge, one arrives at rejecting an attribute any knowledge has, just as the debate about the universalism of knowledge does, and argues until today about the lacking power to reject knowledge, instead of rejecting it via simply critiquing this knowledge. The notion of a scientific universalism is nothing more but the simple statement that knowledge is knowledge independent from where and whom it is created, and it is this attribute that distinguishes scientific knowledge from other exclusive forms of knowledge that invites to argue about what this knowledge tells us what it knows, just as any theories, including those about the scientific universal-

---

[33] Some examples can be found in for example: M. Kuhn, H. Vessuri, *Contributions to Alternative Concepts of Knowledge*, Ibidem Stuttgart, 2016

[34] M. Nieto Olarte, The European Comprehension of the World: Early Modern Science and Eurocentrism, in: M. Kuhn, *Notes on a Critique of how the Social Sciences Think about the World's Social*, Ibidem Stuttgart, 2016

ism, does. It is also the case until today, that theories created in the "West" use the epistemological notion of a universalism of science to disguise the academic power means science has and uses to today, such as positions, money and all that kind of things behind this simple epistemological notion, however, this does not protect these theories against critiquing them—unless one bends down in front of these academic power means. Thus, opposing one of the very rare achievements of the European social sciences, their claim for being provable theories, last but not least in the notion of a universalism of theories, and instead of critiquing their views opposing them of having until today *"a unique and superior kind of knowledge"*, opposes the only achievement these European sciences created and paves the ways for replacing this achievement by the revitalization of the all sorts of non-disputable obscurantisms.

Global social thought under the regime of social sciences knows a radicalization of interpreting global social thought as nationally constructed theories and creates spatiologically constructed social thought—presenting nationally constructed prejudices as science and to do so, sacrifices the only merit the social sciences gained through their critique of the classical philosophies, that is, that scientific thinking is aiming at knowledge, liberated from teleological thinking as thinking through any obscure ex ante pre-assumptions.

... and anti-scientificy to practice global social sciences

The re-discovery of religious thinking, the emergence of sciences like the Islam social sciences and all the celebrated detections of indigenous concepts of knowledge rightly claim to not be less mystic than any well acknowledged ways of social science theorizing creating national prejudices as science, which is as pre-assumptive as any mystic thinking that departs and arrives from its believes— as the social sciences do with their pre-assumptive modelled thought spread across the world as social thought of what they call "modernity", consequently ending up in global social thought as the many prejudiced theories of and about the many "wheres".

One feels attempted to consider it as the impact of a certain place, in which theories are created, that advocates ant-scientism as the means of global social science thinking—and it is, as the way of arguing shows, the social thought in these very places, in which in fact theories can be developed, that consider the nature of science as the enemy of creating a multiplicity of social thought, constructed to represent an authentic unique national view on a local social. It is again social thought from nowhere else but from the imperial world, that, finally, discovers scientism as the major obstacle for the *"universal universalization"* of social sciences, consisting of a multiplicity of spatiologically constructed national knowledges representing national identities:

> "Scientism has been the most subtle mode of ideological justification of the powerful."[35]

Celebrating the nostalgic reanimation of anti-science as means for the global construction of contextualized, spatiological prejudiced theories completes the global reign of an approach to science that distinguishes the world of knowledge into knowledge islands and their clandestine and exclusive ways of theorizing about them. This, the anti-scientism of all the variations of "local" knowledge is a well-received opposition against the "universal" way of theorizing within this very concept of social science thinking, since it frankly articulates the ant-scientism of an approach to social thought and advocates as it practices social thought as a battle against science for the sake of imperial thought.

A final note for those who believe that local knowledge and local social science theorizing was an opposition against "Western" knowledge: Accusing science as the weapon of the *"powerful"*, arguing against *"scientism"*, one should know that *this* is how Wallerstein justifies his *"great moral enterprise of humanity"* only thinkers have, who are troubled by the world power's headache, ruling the world with the support of whom they rule and to search the *"global values"* of mankind, solving this problem for shared "global values", finally not only justifying global military actions as serving these values, but allowing to see wars welcomed as a mission on behalf of these moral values of humanity, welcomed by those who are attacked on behalf of them. He certainly knows why scientism disturbs this *"great moral enterprise of humanity"* of a moral imperial thinker seeking "global values". What those who

---

[35] Wallerstein, I.,(2006) *European Universalism, The Rhetoric of Power,* The New Press, New York p 77

84   Chapter A: The world's social in social science thinking

sympathize with this anti-scientism, because it invites them to contribute any moral view as a contribution to global social science thought might not notice, is that the liberation of social thought from what knowledge makes science, is that this anti-science—sciences also disempowers "humanity" from one of its most powerful weapons. Considerations, which painfully argue about how to carry out wars with the invitation of their victims do not need to rely on any theory, since they obviously trust that their military means will ever make them the "powerful" winner of their wars and are therefore only bothered with their moral headaches, they so generously commission to the worlds thinkers as *"the great moral enterprise of humanity"*, a headache only discreet imperial thinkers can be bothered with.

From patriotic to imperial social science thinking

"Nations matter"[36], under this book title we are told by another US science guru, "one of the most respected social scientists in the world", that "Nationalism is not a moral mistake", a credo this contemporary social science guru reminds the social sciences of, not surprisingly a US scientists, *the* imperial state that established the world of nation states as the globalized[37] rationale of mankind, a scientist who found scientific evangelists in Europe, and who seeks answers for their contemporary concerns about nation states in a globalized world:

> "The question then is how can states for their part win back political meta power qua states in relation to global business actors, in order to impose a cosmopolitan regime that not only encompasses political freedom, global justice, secure social order and ecological sustainability, but revitalizes national sovereignty?"[38]

Global social science thinking, looking at what is going on in the world and to detect the problem, *"how can states for their part win back political meta power* qua *states"* and *to* generously ab-

---

[36] C. Calhoun, *Nations Matter, Culture, History, and the Cosmopolitan Dream*, 2007 Routledge New York

[37] This, the globalisation of the nation state rationale as the globalised way of humans living, is the substance of what is called globalisation. .

[38] Beck, U., http://www.ulrichbeck.net-build.net/index.php?page=cosmopolitan

stract from all the most sovereign executions of political power making the contemporary world a world of wars, is an accommodation of social science thinking to the imperial concerns of nation state power, social science only allow themselves to make when they theorize beyond the nation social. In global social thought social science thinking forgets about all their ideals about communities serving after all the individuals and is concerned about nothing else but the imperial power in front of other "global players", the *"global business actors"*. Global social thought under the regime of social sciences requires an interpretation of the concerns of social sciences that liberates social science theorizing from all their ideals otherwise governing their theorizing interpreting the society, community or whatever social entity as a service for humans and get directly to the point their whole thinking is genuinely about: The—nonetheless, invented—concerns of nation states. They are invented concerns, since it is one of these odd observations only social scientists can make with their discreet glorifications of nation states and all their social missions, to conclude from the fact that nation states compete with their national resources about serving the growth of global capital, that serving capital was not serving the growth of capital from which the nation states, at least the winners of this global competition, not only benefit, but that it is these nation states which set into force and supervise for the sake of their own economic and political power this growth of capital as the top priority of their policy agendas, to conclude from that fact they take the economic means for their power from the growth of capital, that this was a loss of "national sovereignty". It needs obviously the imagination, that nation state power must be so totalitarian, that already the fact that nation states gain the economic means by benefitting from the growth of capital they therefore support with all their political powers means, to interpret the fruitful synergy between global capital and nation states as undermining their power.

Nonetheless, social science thinking remains social science thinking and does not dissolve scientific thinking into any mere political reasoning about the politics of nation states. Concerned about the imperial power of nation states, unlike politicians, social sciences are not concerned about the imperial power of a particular nation state, the imperial concerns of any individual country, but about the imperial power of the nation state as such, the ideas they have about nation states and invent a loss of sovereignty for the nation state as such, generously abstracting from the real power the world of nation states execute. To interpret the fact that nation states measure their economic power against the extent to which the global capital considers as a fruitful service of nation states for the global growth of capital as a loss of national sovereignty (Beck) and of the imperial political power of nation states can only come to the mind of social science thinkers who dream of a world consisting of a "*cosmopolitan regime*" of nation states, an "international community", free of any considerations about serving the needs of global capital, social scientists cannot realize that this service serves the nation states themselves and instead advocate a "*cosmopolitan regime*" of nation states, serving only the imagined aims only social science thinkers want to believe nation states have. And this "*cosmopolitan regime*" is aiming at what?

> "The more cosmopolitan our political structures and activities, the more successful they will be in promoting national interests, and the greater our individual power in this global age will be."[39]

The dream of social scientists about a world ruled by really sovereign nation states suggests to accommodate social science thinking to a new view on the mission of nation states "in the global age" they call *"cosmopolitan thinking"*. Cosmopolitan thinking is a view on the *"global age"* that advocates to promote national interests in the "meta games of power", and that suggests to consider the promotion of nation interests, in other words, that suggests to consider the promotion of the imperial power of nation states in the *"'meta game' of power in the area of globalization" (ibid)*, as a way towards *"greater individual power in this global age"*. Promoting the imperial power of nation states means promoting the individual power of citizens?

> "My second point involves the critique of methodological nationalism in the social sciences. My thesis is: 'the zombie science' of the national that thinks and researches in the categories of international trade, international dialogue, national sovereignty, national communities, the 'nation-state' and so forth, is a 'science of the unreal'. This 'national sociology' is beset by a failure to recognize—let alone research—the extent to which existing transnational modes of living, trans-migrants, global elites, supranational organizations and dynamics

---

[39] *Ibid*

of the world risk society determine the relations within and between nation-state repositories of power'."⁴⁰

To realize how "*the* dynamics of the world risk society determine the relations within and between nation-state repositories of power'" imperial thinking, concerned about "the relations within and between nation-state repositories of power" suggests

> "the idea of '(enforced) cosmopolitanization' to describe the transition from the first to the second age of modernity: cosmopolitanization is a non-linear, dialectic process in which the universal and particular, the similar and the dissimilar, the global and the local are to be conceived, not as cultural polarities but as interconnected and reciprocally interpenetrating principles'."⁴¹

Not only Beck rightly observes that social sciences are indeed not only thinking about isolated nation states as if an outside world does not exist and do this indeed though the outside world "determine(s) *the relations within and between nation-state repositories of power'.*"

Caught by the categorical impossibilities to think about the social other than through the nation state rationale, detecting the *'epochal disillusion'* (ibid) of social science theorizing, a disillusion that simply omitted thinking about the global social as a topic for theorizing, caught by the very way of social science thinking in idealized nation state rationales, a critique that *"the national outlook is consistently identified as a barrier to the effective pursuit of states"(ibid),* a critique that can only phrase the opposition against the national outlook as "*a barrier to the effective pursuit of states*" opposes the national outlook from the very national outlook and, consequently, ends up discussing the effects of the outside world on the nation state socials as a matter of *"the repositories of power",* an outlook that not itself only practices the very methodological nationalism just as the critiqued "zombie science", and therefore arrives at theorizing about the world through the *imperial* rationales through which nation states look on the world of nation state socials and to present this view through the rational of imperial power as giving the individuals more "power".

To put in other words: Beck argues that theorizing that only looks inside nation states is a barrier, but, a barrier for what?

---

40   Ibid
41   Ibid

Thanks to social science thinking, preferably sociological thinking, the critique of thinking in the "national outlook" detects this national outlook, not as a critique of this national outlook but as a "barrier *to the effective pursuit of states"*. Not looking at the nation states beyond is a "zombie science", since it ignores the rest of the world as barrier for what? Not as a barrier to understand how the international affairs beyond an individual nation state affect people's life within the nation states and not as a barrier to understand how their lives are used for the international affairs of the nation states battles about the world power, but as a barrier for the pursuit of the power of states—how could one better apply the very opposed national view to looking at the global social through the national view and to thus arrive at promoting the imperial view national state rationales have on the global social *and* to interpret this imperial view, the effective pursuit of states in the "*global* age", as to the fulfil the mission social science thinkers believe these states have, even in their effective pursuit of their global imperial power.

Indeed, thinking about the social, social sciences ever combine a national realism with a national idealism: For imperial thinkers promoting the pursuit of nation state power in the *"meta game' of power in the area of globalization"* among all the global nation states *and by* promoting their power, imperial thinkers not only know nation states all struggling to increasing their nation state power, but that with the increasing power of nation states *"the greater our individual power in this global age will be."* (ibid)

Needless to say, which national interests of which "where" truly "imprint" social thought to global thinkers, after they praised the objectives of nation state as to increase the individual freedom of the inhabitants of the nation state social, or to say it in the terms of, cosmopolitan thinking, the *"individual power in this global age"*. *Power* is obviously *the* category of cosmopolitan thinking, not only when it reflects about the power of nation states but when it reflects about people in the "global age", the old idea of *freedom* must be the notion of the old zombie sciences.

It must be the art of social science thinking being able not only to present the battle among nation states about their economic power as a loss of sovereignty. Social sciences were no social sciences if they would not be able to discuss the battles about the power among nation states about power as a problem for the power, not of any nation state, but of the nation state as such. Looking at the world facing the power battles among nation states, social sciences not only are concerned about the ideas they have about what a nation state are, but they are able to imagine that the global battle among the powerful nation states about global power, that this very battle not only increases the global power of all nation states, but that the imagined increase of sovereign power of nation states results in the greater individual power of the world's citizens, those very citizens, all nation states use as their means to fight these global power battles. Remarkably, while social science reflections about the power of nation states inside nation states are presented as serving the interests of its citizens, global social science theorizing, in our example called "cosmopolitan thinking" detects the imperialism of nation state as serving their citizens—serving their "individual power". In the imperial version of social science thinking social science thinking accommodates the objectives of nation state citizens to the objectives of imperial nation states and present the objectives of citizens in being interested in nothing else but— power. Hard to decide what is more appalling, the frivolity or brutality of such thought, identifying the imperial substance of nation states, the power over other nation states with what citizens must aim at, power, a new view of the social sciences on the "global age" presented as the "cosmopolitan" interpretation of the world's social, a view liberated from the national views of the old "zombie sciences".

A leading US scholar demonstrates what it means to practice hegemonic social thought, a way of thinking that overcomes thinking in which *"the national outlook is consistently identified as a barrier to the effective pursuit of states" (Beck, ibid),* illustrating the idea of *"cosmopolitan"* thinking, serving *"national interests"* and by doing this serving the *"individual power".*

Wallerstein, a thinker from that part of the world, where the idea of cosmopolitanism constituted the rationale of the supra-imperial nation state supervising the world's imperialisms, discusses a question this most distinguished social science scholar feels *"goes to the heart of the political and moral structure of the modern world of the modern world-system",* which is how to legitimate global wars among what Beck call the "power repositories", that is, when they not only always store their power means in their power repositories, as they do in Beck's image about nation states, but when they use them:

> "The question—Whose right to intervene?—goes to the heart of the political and moral structure of the modern world of the modern world-system. Intervention is in practice a right appropriated by the strong. But it is right difficult to legitimate and therefore always subject to political and moral challenge.[42]
> ....It is not that there may not be global values. It is rather that we are far away from yet knowing what these values are. Global universal values are not given to us; they are created by us. The human enterprise of creating such values is the great moral enterprise of humanity. This issue before us today is how we may move beyond European universalism—the last perverse justification of the existing world order—to something much more difficult to achieve: a universal universalism, which refuses essentialist characterizations of social reality, historicizes both the universal and the particular, reunifies the so-called scientific and the humanistic into a single epistemology, and permits us to look with highly clinical and quite skeptical eye at all justifications for 'intervention! By the powerful against the weak."[43]

Finding justifications for "interventions"—he talks about the US wars, just to clarify this, not about theoretical interventions—, finding "*a universal universalism*", shared by the whole world,—sorry, wrong, Wallerstein is an idealist sociological thinker so, we have to re-phrase it this way—having a "*quite skeptical eye at all justifications for 'intervention,*" is the mission of global social thought and for this mission he is seeking a "single epistemology", and, by doing this, finding "*global universal values*", this dream of any global empire making their imperial wars a desire of those they attack, doing this is nothing less but fulfilling the "great moral enterprise of mankind", this is how most typically the, no doubt, most critical US scholar bothers the world of science with *the* moral headaches of an imperial mind he discusses as the major mission of global social thought. Thinking about the world's social and about the world's social thought as serving to solve the moral problems only a view has, which appropriates the essential of the rationale of an an imperial world view, and which presents the genuine mission of the imperial view on the world' social as to represent and to care about the values of mankind, only such a view on the world's social can most critically raise his hand and warn the world that "*that we are far away from yet knowing what these values are.*" Only

---

[42] Wallerstein, I., *European Universalism*....,p 29
[43] Ibid .... p 79

thinkers who are always on a mission for mankind, cannot think about the world other than through a global "we", ever embracing in their social thought nothing less but the "we" of the world.

Wallerstein, certainly a pioneer in global hegemonic thinking, not incorporating the state rational of a single nation state, but, as a social science thinker thinking about the world' social, is not at all an exotic exception, but most typically demonstrating how the imperial nation state rationales constitute the imperial modes of social sciences thinking when social science think about the world's social and what social science thought is about that encompasses what Beck critiques as the *"national outlook, a barrier to the effective pursuit of states"* and that advocates imperial thinking as the way of social science thinking in the era of globalization about the world's social, these thinkers call "truly global science".

Nationalism: A service for imperial social science theorizing

Theorizing about and through the perspective of nation state constructs, must have a narcotizing effect, when it advocates imperial thought. While people like Beck present thinking through the imperial power of nation states as a means to increase the power of individuals across the world, calling this imagination "cosmopolitanism" and encourages the social science " to kiss the frog", just as if imperial thinking was liberating a princess from its disguise, other more realistic social science thinkers are concerned about the sociological idea of "cosmopolitanism" and feel that this idea might end up in dissolving the nation state world order, a world order a social scientists like Calhoun admires, thanks to his scientific expertise he has in thinking as praying the nation states as a means for thinking about the world:

> "I have been writing on nationalism since the early 1990s and reading about it much longer:"[44]

Summarizing his knowledge about nationalism, the celebration of ones nationality, he articulates his concern about theorizing about

---

[44] Calhoun, C. (2009), *Nations Matter,....*, p.vii

the imperial concerns of nation states under the notion of "cosmopolitanism":

> "Nationalism matters not least because it has offered such a deeply influential and compelling account of large scale identities and structures in the world—helping people to imagine the world as composed of sovereign nation states. The world has never matched this imagination, but that does not deprive the nationalist imaginary of influence."[45]

> "Nationalism, then, is the use of the category "nation" to organize perceptions of basic humans identities, grouping people together with fellow nationals and distinguishing them from other members of nations. It is influential as a way of helping to produce solidarity within national categories, as a way of determining how specific groups should be treated (for example, in terms of voting rights or visas and passports) and as a way of seeing the world as a whole. We see this representation in the different colored territories on globes and maps, and in the organization of the United Nations"[46]

> "We need to respect the importance of belonging to nations and other groupings of human beings smaller than humanity as a whole. We need to understand that such belonging does different sorts of work for different people—inspires some, protects some, consoles some, as well as makes political opportunities for some."[47]

One wonders, what is more striking about this social science thinker, his style of phrasing things as a giving advice like a science priest if not to "humanity as a whole", but at least to the whole "we" of the social sciences on behalf of which he seems to give his warnings or his paranoid concerns about the danger that the very sociological idea any sociologist across the world shares might disappear, which is that "humanity as a whole" seeks nothing more but what any sociologist finds the most essential desire of any humans, that is their belongingness, and that it is the nation state that offers for all those who seek belongingness to satisfy their desire for belongingness with their nationalism. For non-global sociologically thinking, for more ordinary people his concern is to cultivate and to defend something smaller, something more handy than thinking about the "humanity as a whole", something smaller than "cosmopolitanism", and yes, this smaller version of the sociological idea of belongingness thankfully exists, it is something like nationalism

---

[45] Ibid, p.8
[46] Ibid, p.3
[47] Ibid, p.9

and it is precisely the concept of nation states which is the appropriate scale and category of thinking for nationalist thinkers to understand—a world of nation states. Nationalism, unlike that confusing concept of cosmopolitanism, must be seen a great invention to understand the who is who in a world of nation states.

Anyway, nationalism, is a great sociological construct to understand the belongingness to nation states, which already thanks to the limited size nationalism embraces, less than "humanity as a whole", easily allowing people to identify where they belong to, people he considers intelligent enough to know about their "belongingness" thanks to "*colored territories*". If these are not enough reasons "*to respect the importance of belonging to nations*", one may just imagine nationalists, seeking their belongingness to their home nation state in a world that was without—the small—nation states, only the huge "humanity as a whole"! What a great idea having passports "*as a way of seeing the world as a whole*". Otherwise, who are "we" in a world of many national we's?

> "The constitution of nations—...—is one of the pivotal features of the modern era. It is part of the organization of political participation and loyalty, of culture and identity, of the way history is thought and the way wars are fought."[48]

Imagine a world of a "humanity as a whole", a world without nation states, how could humans know to which political entity they must prove their loyalty? And, how could "humanity as a whole" know whom to shoot on "in a war"?

> "The most basic meaning of nationalism is the use of this way of categorizing human populations, both as a way of looking at the world as a whole and as a way of establishing group identity from within....
>
> The two sides come together in ideas about who properly belongs together in a society, and in arguments that members have moral obligations to the nation as a whole—perhaps even to kill on its behalf or die for it in a war"[49]

This imperial thinker, advocating imperial social science thinking via nationalism as opposed to cosmopolitanism, knows what the whole talk about nation states as a means for people "that organizes people's sense of belongingness in the world" is all about. It is the service for the sense of belongingness that knows how to dis-

---

[48] Ibid, p.49
[49] Ibid, p.39

tinguish the world into friends and enemies and to know whom to kill and for which nation state to die. Fighting wars without nation states, fighting wars without knowing who is the enemy? No, impossible, nation states are a "pivotal" organization allowing us to know precisely, whom to shoot and whom not.

Social science thinking about nationalism seemingly has the effect blurring the senses making social science thinkers a case of particular sociological paranoia, the paranoia of a non-structured live, that makes sociological thinkers to discuss killing and dying as a service for their "sense of belongingness". Sociological thinkers are so deeply concerned about imagining the subalterns of nation states without their ruling power, imagining the creatures of nation states without its creator, that they admit that the rationale of nation states might be also be responsible for all kind of violence, poverty, racism, "the way wars are fought", and that the rationale of nation states from time to time requires to sacrifice the life of its inhabitants for the nation states imperial wars, but essentially even these wars must after all be also considered as organizing orders, structures and identities. Imagine if passport holders of another nation state would just mention their names without indicating any nationality, it would be impossible to imagine "*how specific groups should be treated*", which are not "determined" as nationals and which might be even without nation states. How could one see the world without nationalism "*as a way of seeing the world as a whole*" if the globe was not a globe of nation states "*we see ...in the different colored territories on globes and maps*", thanks to the existence of nation states.

Considering the typical sociological passion, arguing about nationalism imagined as providing nothing but "order", "structure", "identity", concluded from off-thinking the very ordered subjects without their ordering power, one tends to overlook the concept of organization thinkers must have in their mind, brashly presenting nation states as organizing human life. Presenting the life of nation state citizens that is setting into force to practice life as the vice versa dependency of achieving ones interests by preventing the others to achieve his, at the same needed to strive for ones owns, presenting this nation state social life everybody against everybody, not even mentioning the organization of the economic life in a market economy with its economic battles among all economic subjects, presenting this nation state social life that articulates in any law how to arrange interests that essentially exclude each other, presenting this forced battle among nation state creatures as a way of organizing human live, is only thinkable for a mind, that, indeed, again must imagine all these nation state humans without the ruling power of nation states, nation states which for obvious reasons need above all nothing more but a monopoly on power in front of their citizens, the very nation state who forces its citizens into this everybody against everybody.

May this be considered as a well-organized society, imagining the horror vision of a world of nation states creatures without nation states and arguing from there that nation states are undeniable the only guaranty to identify the world of nation states as a world of nation states, is no tautological logic, but most distinguished sociological thinking among social scientists most seriously concerned about the question if thinking through the imperial view of nation states is not eroded by a notion, cosmopolitanism, that in the eyes of such nation state freaks insinuates for fundamentalists sociological thinkers and their worries about a world with no "structures" the dooming end of nation states and, to make it worse, the end of social science thinking:

> "Nationalism is easily underestimated.....Analysts focus on eruptions of violence, waves of racial or ethnic discrimination, and mass social movements. They fail to see the everyday nationalism that organizes people's sense of belonging that leads historians to organize history as stories in or of nations and social scientists to approach comparative research with data sets in which the units are almost always nations."[50]

Worse than imagining all the nation state creatures without its ruling creator, and to conclude that nation states are the best organizers of a life of the everybody against everybody, only the nation states set into force, and worse than imagining that nationalism would lose its orientation of belongingness if there was nationalism without nation states, worse than all these dis-orders only social science thinkers make up to pray the existence of nation states, worse than this is for social science thinkers, to imagine imperial social science thinking without the nation state, as their unit of analysis, as their imperial perspective to think about the world of

---

[50] Ibid, p 27

nation states and as their theoretical guidance ordering thinking about the who is who in the imperial world order.

This is at least a very correct observation that nation states "lead historians" and the whole social science theorizing, though it is a less worrying imagination, to imagine a world without both.

...thought back by alternative imperial social science models

There seems to be nothing in the world of social sciences, may this be not only advocating imperial thinking, like Wallerstein, Beck and Calhoun, that does receive a critical response—with an alternative of the same. While Wallerstein constructs the world of science as a mission of an invented global we, alternative imperial thinkers advocate and practice a counter model of an alternative imperial view, consisting of a battle between two global "wheres", a "Northern" versus a "Southern" body of theories.

Thinking through an idealized state view on the social, founding the reflections of social sciences to reflect on the world's nation state social, is, thanks to the universalization of the social science approach to social thought, no longer the privilege of social science thinkers among social sciences in the imperial world. Once the formers colonies found nothing more desirable but imitating the very society system that oppressed them as colonies and imitated the otherwise rather discriminated but critiqued "North", the former colonized countries not only imitated the society system of the former oppressors but also their science system and are unhappy that the theories from the "North" not coincidentally correspond to the society system the former colonies imitated and dominate global theorizing, arguing that the "South" is lacking its own authentic nation state view and as a consequence suggest explicitly patriotic views on the global social as the new mission of alternative social science thinking.

Social science thinking in the science world calling themselves "peripheries"—thus revealing that their opposition to the sciences in the imperial world is a gradual difference regarding their lacking possibilities to develop their sciences towards the same impact of knowledge—have learned their lessons from the colonizers in thinking about the world's social from the view of an idealized nation state perspective on the word's social creating social thought about and through their "places".

Reflecting on social sciences in the "Third World" countries, they articulate their following concerns:

> "If we consider the parallels between economic dependence and academic dependence we may define the latter as a condition in which the social sciences of certain countries are conditioned by the development and growth of the social sciences of other countries to which the former is subjugated. The relations of interdependence between two or more social science communities, and between these and the global transactions in the social sciences, assumes the form of dependency when some social science communities (those located in the social science powers) can expand according to certain criteria of development and progress, while other social science communities (those in the Third World, for example) can only do this as a reflection of that expansion, which can have mixed effects (positive and negative) on their development according to the same criteria."[51]

With their odd associative construction of "parallels" between the economic relations between the imperial world and the developing countries and the relations between social sciences in the "peripheries" and the "centers",—a comparison in which they present the conflicting economic relations as the rather harmless relation of "dependence", thus extinguishing the essential conflictual relations of exclusive and opposing interests—, social science thinkers in this "peripheral" world demonstrate their mastership in thinking in idealized nation state rationales and prove that their way of thinking is very much thinking in the centers of social science thinking, while discussing the contributions to knowledge from thinkers in the third world as a concern for the "Third World's"—nation states.

While Wallerstein's grumbled about a monstrous "universal universalism" [52], thinking about how to find nothing less but a global "we", and while critical thinkers like Adorno are concerned about the morality of the Germans, social science thinkers from the "Third World" ever also construct the "Third World" as another global counter-"we" of national subjects, in which they do not want to know anything about the many conflicting interests between all the subjects within these nation states, constituting this "we " in the former colonized world, now also nation states, may this be

---

[51] Dos Santos, Theotonio, (1970), *The Structure of Dependence*, American Economic Review, p 603
[52] Wallerstein, I., (2006) *European Universalism....*, p 79

conflicts between the nation states of the "South", may this be between the citizens within these new nation states and between the nation states and their citizens. They master the social science manor gathering subjects only they suppose share a dependency from another global "we", generously ignoring all kind of dependencies these subject have among themselves within these nation states, as among the nation state within the "developing" world. Social sciences in the "developing" countries also appreciate to create the ideal of a "we". They unify not only nation states and gather them under the notion of "dependency", easily ignoring their wars against each other. They also practice social sciences thinking in and across these countries as a unity only they see, a unit, they construct negatively as sharing a victimization of the what they therefore reversely gather under the notion of the "Western science", not opposing a social science approach, but as theories of social sciences characterized as the opposed group of nationally constructed entities, the "scientific centers", they also present as the revers false unit as their "developing" social science unit from the "peripheries", a relation between entities, constituting the very meaning of the category "scientific dependence", under which they are gathered.

It requires a very non peripheral determined social science view on the sphere of social science, to accommodate thinking about the competitive academia in the developing countries and to call this competing academia a "social science community", those in the "third world", and to present their contribution to knowledge as a problem, not for the world's knowledge, but for these multinationally constructed imagined entity, the "South", the imagined social science "community" of the "Third world", opposed to the accordingly imagined "Western sciences".

From there on—not only—academics from the "Third World", as the social science generally do, they also count the nationalities of thinkers, whose contents of thought do not only not really matter. They all count independent from what they say about these countries as about the "centers" as welcomed contributions to the social sciences of the "peripheries". The view on social sciences as a nationally identified activity, not just a few scientific thinkers in the developing countries share, thinking about social sciences even radicalizes the view on science as a matter of the position—ever idealized—nation states have on science, just as if the view of national policies and the view of thinkers on the world of social sciences was the same, and engage themselves in discussing the international status of their "Third world" scientists representing simply thanks to their nationality nation states or nation state entities in what they construct in the social sciences minds as a global battle about an "expansion" of science. For social science thinking in the developing world, creating knowledge about the world's social and to strengthening the "Third World's" nation states position in global social thought consisting of a multiplicity of spatiological knowledges, is in their imagined battle between an imagined "center" and a "periphery", seemingly the same. Social sciences from the developing countries compliment the view on social sciences from an imperial view on the world' social and construct an alternative global imperial counter "we", gathering social thought under an alternatively constructed politically "southern" social science entities, entities only social science theorizing can construct that identifies global social thought as matter of the competition among a multiplicity of nationally constructed knowledge bodies unified in a global battle among imperial knowledge bodies arguing not about which knowledge allows us to understand this world, but which knowledge "dominates" thinking about the world—the inequalities in knowledge "expansions".

...critiquing an unequal knowledge imperialism

Social science theorizing is able to oppose and to advocate difference and identity in the same thought.

"Inequalities", the complaint about difference, is an example for a fashionable[53] way of social science theorizing, which gathers critical discourses across the world after the world has been subordinated under the same economic and political system, which at the same time insist on "difference" as the main essential category of all the global ant-identity freaks, whenever they compare any topic across nation states, discourses, which argue under the notion of "inequalities" against difference, not only about unequal science powers between "centers and peripheries", but under the same notion against the global "inequalities" in the distribution of power and capital.

---

53   See Section D in this book

The image these social science discourses have about the world is relatively simple: Firstly: The world is a world of nation states, this is a "fact". Secondly: All nation states must insist on their uniqueness against others, this is the nature of nationalistic thinking. Thirdly: All nation states must have equal power in the battle about ruling the world, this is a neat ideal about a fair imperialism.[54]

Imperial thought are thought constructed through the concerns of nation states about their ruling power in the global battles among nation states and this is thought that both creates false theories about the global social and is part of a discourse among nation states using their people as a means for their battles about global political power. It is social thought which is anything else but an opposition against imperialism. Wallerstein, Calhoun and Beck are examples from the imperial world advocating imperial thinking.

"Inequality", the negative phrase of what is otherwise called difference, is not the same as difference, because "inequality" is more than the observation of difference; it is a critique and it is a critique that could not be more typical for social science thinking. It is a critique that does not argue about what and why it opposes the social world as it is, but it opposes that the social world is not what it is supposed to be, according to what social sciences thinking wishes it should be: equal.

With this idea in mind, social science thinking detects a world that consists of inequalities everywhere, translating any difference into failing to be what the social science dream is, and it is this failure, that is the whole critique.

Hence, instead of thinking about any such difference, what the difference is, what causes such differences, thinking about difference is no topic for this critical idealism, since inequality is already

---

[54] With regard to sociological reflections about the global social sciences, discussing scientific power relations as a matter of "scientific power" illustrated such thinking that does not want to oppose any power relations in sciences, but eagerly discusses their "inequalities", just as if any scientific power would aim at equal powers, abolishing itself. See also, M. Kuhn, 'Academic Dependence': The World Social Science Arena—a Battlefield among Parochial Social Thought, in: M. Kuhn, K. Okamoto, *Spatial Social Thought, Local Knowledge in Global Science Encounters,* Ibidem, Stuttgart 2013

all what needs to be thought and said. Any substance of any difference does not inspire this thinking to think about its substance but is taken as facts that proves—"inequalities".

Passionately finding through this dreamy view a world full of difference accused to not match with the ideal of the social science thinkers, difference heavily told by the social science thinker that it should not be there, is entirely ignorant not only about what the difference is about, but therefore never comes across that differences might turn out to be not only difference, but contradicting things.

There are inequalities in education, social relations, income, gender and among nation states and anywhere any difference is made an inequality.[55]

For the view that sees the world full of "inequalities" it does no longer come to this dreamy mind, that the difference between the developing and the imperial countries is not simple a difference, both are not just gradual variations of the same, but one, the imperial countries, are the reason for why developing countries are ever

---

[55] Most recently the French economist Piketti has become famous for warning capitalism, that flourishes worldwide with ever growing rates of dividends, last but not least thanks to the lowered salaries of the employees creating this wealth, that this booming business world is at risk, because it causes so many—inequalities. All the tables filling a book with more than 600 pages document nothing but inequalities from the 18th century until the 21first, all threatening capitalism with its self-made death equalities, unless, as this thinker against inequalities recommends, the nation state cares for income inequalities, thus keeps a middle class with higher incomes as the working class as the basis of the nation state. The income of class of capitalist seemingly fulfils the needs for equal inequalities in this dooming scenario. Besides, this economist has collected so many data, proves for inequalities, to advocates an unequal distribution of wealth, that he does not want to notice from all his tables, that capitalism has nothing to distribute, as Marx already argued numerous times against all those distribution ideals. Who owns what, is not only already decided via the antagonistic economic interests fixed with the power of nation states in the not only different but exclusive substance of different private property rights, not only before any wealth is produced, but as the condition for any production; a view in his table could show him, that the political regulations about who owns what, the exclusion from any wealth before any production and, hence, from the wealth to be produced is the sine qua non condition for investors, to invest their private properties into any creation of wealth, so that all wealth is already distributed, not unequally, but exclusive. So, no worries about any disappearance of "inequalities". Piketti, T., *Capital in the Twenty-First Century*, the Belknap Press of Harvard University Press, Cambridge 2014

developing countries. Thinking about the difference of both would reveal that they are not only different, but that there nation state rationales are complementary enemies. Such things, like exclusive interests never come to the mind of the inequality freaks, since there world view is the view of dreaming of a world of harmony, accusing the world to not be harmonious.

Arguing for a world of equalities, the world part tout does not aim at, inequality freaks become quite inharmonious and reactionary, oppose the world for failing to coincide with their imagined world image, warning the world of inequalities that it would collapse if it does not reduce ore abolish inequalities.

Inequality social science thinkers, applying their dreamy world view to the global world of nation states, cannot think of opposing imperialism, the antagonistic interests political powers decide via wars, but argue about the inequalities of political power and inevitably end up in advocating an alternative imperialism, the buffoonery idea of an imperialism in which the world nation states consist of equal imperialist nation states. The imperial world must shiver accused failing to build a world of a fair imperialism.

Hence theories, discussing contradictory things, preferably Marx, are critiqued for ignoring—inequalities. Writing a theory about what capitalism is, an economic system that existed in the colonizing countries, the imperial countries did not apply those parts of the world where they simply robbed the wealth, what is making them colonies, such a theory is critiqued for not seeing colonies as documenting a world consisting of—inequalities; a remarkable postcolonial way downplaying the antagonistic relations between the colonies and the colonizes as a gradual difference of the same, an idea, ironically and preferably fabricated from social scientists in the former colonized world, an idea that only comes to minds, who cannot think about the colonial world other than through the perspective of nation states the postcolonial imperial countries, namely the US, established, making the world a world of nation states, constructed along the US concept of nation state.

The world's nation states serving the world's mankind

The one remaining question, why social sciences consider reflecting about and through the rationale of nation states ending up in thinking about the imperial concerns of nation state, as the nature of social thought and for doing so, sacrificing scientism for the sake of advocating "local" views in globalizing sciences, social sciences themselves answer strikingly simple:

> "To build a truly global sociology, I first have to make clear my standpoint as a Japanese sociologist and as a member of the sovereign people of Japanese Society."[56]

Social sciences scholars conclude from that fact that nation states, in which they were born, define them, once born, as a citizen under their national law, that their standpoint is the standpoint of the very nation state that defined them as their national subject.

Alternative social sciences devote themselves a wider mission, not thinking about and through an individual nation state, but through the concerns of *the idealization* nation states, more precisely from the perspective of an entity, the global South, an entity that only exists in the mind of critical social scientists:

> "The purpose of this book is to propose a new path for social theory that will help social science to serve democratic purposes on a world scale.... I use the term "southern" not to name a sharply bounded category of states or societies, but to emphasize relations...between intellectuals and institutions in the metropole and those in the world periphery"[57]

A two hundred years' experience seemingly is not enough to create any hesitations about *the* ideal of the concept of nation state, democracy, though everybody who just opens the eyes could notice that democracy is not more but the simple procedure to empower political elites in nation states—given they find this procedure opportune. How many more example do the idealist of democracy need to interpret elections, ever hold after the political landscape has been cleaned up from all alternative political parties, after democratic parties forbid them to take part in elections, not as the

---

[56] SHOJI Kôkichi, *unpublished paper for the Yokohama ISA World Congress*
[57] Connell, R., (2009) *Southern Theories,* Polity Press, Cambridge, p vii

dictatorship of democrats but as violating their ideal of a true democracy? What happened to all the countries where democracy has been introduced, after the political scenery was cleaned, an opposition suppressed and then people were called to elections, has any of the essential living conditions been changed after the political elites allowed their people to enthrone them via elections? More than 60 years' experience of the post-colonial world consisting of a world of nation states facing an ever increasing level of poverty and violence the world did not see in history before, is not a hint for social sciences to question their mission for the sake of this nation state ideal, in this case, not reasoning a mission of social sciences for an individual nation state, but for the not *"sharply bounded category of states or societies... in the world periphery"*.

And finally, even when social sciences rightly critically detect, that social science thinking is thinking through national "outlooks", there alternatives, opposing thinking in national perspectives and suggesting thinking beyond the nation state views, only knows to replace the national outlook into the nation states social—by the national outlook nation states have on the world' social and arrive at imperial thinking, thinking that looks at the global social through the imperial rationale of nation states or its alternatives from any invented state units like the global "South" or the "peripheries".

And even this, thinking about the world's social through the imperial view nation states have on the world inside and outside their sovereignty for the global battle among nation states about global political and economic power, "truly global" thinkers want to believe that the *"meta game' of power in the area of globalization"* finally serves the "individual power". Obviously, truly global thinkers can no longer think about the social other but in the categories of *power,* when they idealize imperialism as a service of nation states for their citizens.

What makes social sciences such idealistic and critical thinkers for and through the idealized rationale of nation states that cannot be ever irritated by any experience of wars, poverty and violence, the world is facing more than ever before in history? Why does it never come to their mind and detect that their 200 years lasting critique, ever accusing nation states history as failing their mission, that what social science consider as failing their mission might, after all, not at all be their mission? The answer on this question can be found in how disciplinary thinking works, that provides social science thinking with the categorical essentials through which social sciences theorize.

Chapter B
Categorical essentials
of disciplinary thinking[58]

Social thought under the global regime of social sciences is knowledge about spatially confined socials constructed through the particular view of nation state theorizing, finally scientific knowledge that sacrifices the epistemological essential of the very social sciences for creating global knowledge as a collection of national prejudices.

Social sciences knowledge is disciplinary knowledge. Contributing social science theories to global social thought is contributing knowledge that not only contributes to disciplinary knowledge but it is knowledge, constructed through the categories of disciplinary thinking. Thinking in the social sciences is disciplinary thinking about the social; it might well think about nationally confined social entities, however, there are no theories in disciplinary thinking constructed through disciplinary confined spatial categories: There might be very well disciplinary theories about any nationally confined social, but they are constructed through categories and meta—theories, which represent thinking about the social that is not restricted to any spatially constructed categories.

The theoretical foundations of disciplinary thinking are not constituted by spatially confined categories.[59] There is disciplinary

---

[58] Unlike the distinction among the social sciences into disciplines, the distinction between social science and natural sciences *is* a matter of the object of theorizing. Their division, however, is due to the view the state gained on both sciences to separate them: It is the very same instrumentalism of this utilitarian view on the society and the nature, using both for the purposes of the privates, that results in rationale natural sciences and in the idealization of the society of privates presenting the society as a means for the privates in their categorical essentials, the ideas about the homo economicus and politicus, as a means for pursuing the interests of the private property owners as the nature as a means for pursuing these interests.

[59] There are, of course, debates about the categorical foundations of disciplinary thinking across the world, but the discourses about the categorical foundations of theorizing within disciplines are aiming at adjusting the categories of the disciplines, not substituting disciplinary thinking by national

theorizing in and about a national social, but there is no nationally constructed disciplinary theorizing. Sociological thinking, for example, is a way of disciplinary thinking about the social, it is about national socials, but there is no multiplicity of sociologies coinciding with the multiplicity of national socials. There may be though different interpretations of what the categories founding sociological thinking are, however, those different interpretations of the categorical apparatus of any individual discipline remain interpretations of an individual discipline.

Nevertheless, disciplinary thinking participates in a global knowledge "production" by adding knowledge to disciplinary knowledge with knowledge that is not only about nation state socials but that must be constructed through a nationally prejudiced view on any disciplinary topic, thus violating the categorical basis constituting disciplinary thinking and using these disciplinary categories to contribute knowledge to the disciplinary knowledge.

Justifying the violation of disciplinary thinking with the need to construct nationally presupposed theories for contributing to global social thought in the social sciences and doing this with the epistemological foundations of disciplinary theories is an odd paradox[60] of global social thought under the regime of the social sciences. Detecting, consequently, scientism as suppressing the articulation of theories that violate disciplinary thinking and advocating to replace scientific thinking by moral disputes about "cultural" values and, finally, reasoning a dispute about values as the only way to arrive at global *scientific* thought is another one. How does this go together, spatially constructed patriotisms and disciplinary knowledge; justifying the violation of the rules of social science theorizing with these rules and opposing scientific thinking with the means of scientific thinking? Social science theorizing apparently can combine both while thinking about the world's social.

---

      thinking. There are attempts for creating—for example—an Arab sociology, but they are still arguing for and within disciplinary thinking, here sociology.

[60] Social sciences reflections are struggling with this paradox. As an example see: Towfigh, E, Ahmada, S. (2013) How to Overcome "Oriental" Sociology, in: Kuhn, M., Okamoto, K., Spatial Social Thought, Local Knowledge in Global Science Encounters, p. 313ff, Ibidem, Stuttgart

The cognitive architecture of disciplinary thinking[61]

Social sciences are differentiated in a diversity of exclusive ways of disciplinary thinking by offering a multiplicity of disciplinary explanations about the same social, all claiming to provide theories explaining the same social phenomena.

The plurality of social science theories originates from a multiplicity of disciplinary categorical frameworks for theorizing, chosen not only before but also for theorizing. It is this multiplicity of reflexive frameworks, disciplinary thinking provides for theorizing that is responsible for the plurality of social thought in the social sciences' approach to social thought, a plurality of theoretical presuppositions, through which social thought in the social sciences prefabricate the object of theorizing for a particular explanatory model and to thus arrive at a multiplicity of coexisting and exclusive theories. The plurality of disciplinary divers explanations for the same phenomena are not only not bothered if one theory explains a phenomenon as a case for economic thinking, thus a matter for thinking about the economy and another theory considers the same phenomenon as a matter for psychological theorizing, a theory that must be found in the mind of humans.

Disciplinary theories do not even very much argue with each other about their diverse theories about the same object of thinking, but rather consider the diversity of their theories as complimentary thought that could be assembled through an inter-disciplinary knowledge creation, as if theories with diverse explanations would arrive at *complementary* knowledge if they assemble their *exclusive* theories.

Disciplinary thinking, subsuming any object of theorizing under the disciplinary explanatory model for theorizing is for disciplinary thinking not the origin of a multiplicity of disciplinary presupposed knowledge, but a necessity that originates from the nature of the social. However, this, reasoning disciplinary thinking, the need for choosing between varieties of presupposed disciplinary frameworks for theorizing, with the nature of their object of social

---

[61] These reflections about disciplinary thinking focus on the classical social science disciplines, that is on anthropology, economics, sociology, political sciences and psychology.

thought, is the pretentious justification of the very disciplinary thinking for the need of this varieties of disciplinary thinking.

It is simply not the case that, as disciplinary thinking itself justifies disciplinary thinking, disciplinary thinking mirrors a division of labor, covering the distinctive elements of the particular way this society model constructs the essentials of humans' life, disciplinary thinking presents as the nature of humans. It is obviously not the case, as the justification of a disciplinary thinking as a division of labor states that the discipline of economic thinking thinks about the economy, sociology about the civil society, political theory about politics and psychology about the mind of individuals or groups of individuals. All disciplines are not born via a particular subject they think about, but by a particular way of reflecting about the same subject. There are psychological theories about politics, sociological theories about the economy and economic theories about politics and so forth; and there are endless combinations of disciplines such as socio-economics, psychology of politics, economics of psychology and so forth and there are endless battles among the disciplines over their definitional power, claiming their disciplinary definition of the same object of thought as its disciplinary nature defining their reflexive "territories". The reason for a multiplicity of theorizing about the social is not a response to the nature of the social, this only justifies only one social science and not the plurality of social sciences, but this plurality of particular perspectives for thinking about the same objects of theorizing, approaches to theorize, only disciplinary thinking constructs, justifying an theoretical framework through which they theorize, presented as a necessity of the social reality.

It is only in the social sciences that social thought interprets the social, economic and political aspects of human activities as a multiplicity of a disaggregated human being, just as if these humans have three different existences. Thinking the political citizen abstracted from his economic actions is the exclusive view of a way of social science theorizing which interprets the different aspects of the same social actions as actions of different humans. In the eyes of social sciences thinking the human of the bourgeois society becomes decomposed into a multiplicity of subjects, each constituting a distinctive theory developed from its particular construct of a nation state human: As a political "citoyen" the nation state human constitutes political science thinking through politically defined aspects, sociological thinking through his social life, his interrelation with other nation state subjects of the same kind, as a subject reproducing his material life he constitutes economic thinking, and through his battles with his own mind and will to master the challenges of this very nation state life, psychological thinking, just as if thinking from these different live perspectives of the same nation state life was not *the same human*.

Unlike social sciences argue, interpreting social science disciplines as a division of labor[62], the distinction of social thought in distinctive views on the social is the result of the particular way social sciences pre-construct their object of theorizing.

It is no doubt the case that human life has distinctive essentials such as the material re-production of humans, an economy, such as a communal body, jointly organizing their communal concerns, a polis, and such as the ways individuals mentally handle their lives, a psychology, as possibly some other such essentials, constituting human life and thus a need for reflecting scientifically on them. However, the fact that human life has these distinctive essentials, is not at all the same as constructing the interrelated components of human life as separated elements, separated into a bunch of disaggregated life instances, justifying the separation of social thought into disciplinary thinking.

And it is also the case than human life in any society has economic, political, social elements that need to be thought about by social thought. Disregarding the question about what issues a society, unlike the nation state society model and its market economy, in which nothing, including the economy, is not politically regulated and ruled and though nobody is able to anticipate the inscruta-

---

[62] It is amazing to see social sciences reflecting about the social sciences, notwithstanding with the way of thinking social sciences invented. Carefully tracing over more than two hundred pages a typically historic description of the emergence of social sciences, of the disciplines and their battles about the scientific topics, without at any point raising the question, why, for instance, social science theorizing creates a differentiation of the classical philosophies into a set of disciplines, or, why political science and sociology argue about what their scientific territories are. None of all the questions the emergence and the historical development of disciplinary thinking so obviously provoke are neither raised nor discussed, just as if the emergence of social sciences was to finally arrive at the nature of thinking about the social. Wallerstein, I., (2001) *Unthinking Social Science, The Limits of Nineteenth-Century Paradigms*, Temple University Press

ble developments of this economy, disregarding the question about what issues a society would need scientific knowledge if the society crafts its life with scientific knowledge, undoubtedly any social thought also beyond the social sciences needs knowledge about the economy, about politics, the social and the individual. It is however a very different thing, to think about any economic phenomenon and to do this with the existing economic knowledge or to subordinate any social phenomena under the explanatory framework of a social science discipline, whatever the topic at stake is. Disciplinary thinking is everything else but using the existing knowledge assisting the thinkers to understand his economic phenomenon. Disciplinary thinking rather applies its theories as a dogmatic interpretation framework to any topic it reflects on, whatever the obvious nature of the topic is: Explaining war as a matter of uncontrolled aggression, education as a an input/output relation of investment or profit as violating the social commitments of some irresponsible citizens are theories resulting from an ex ante choice for a disciplinary framework applied to any social phenomenon and are considered in the social sciences as theories, that contribute serious scientific knowledge. It is indeed something else to think about any social phenomenon using the scientific knowledge a society has and, preoccupied by the believe that any particular disciplinary interpretation framework must allow to understand any topic whatever its nature might be, to impose arbitrarily any disciplinary interpretation framework on any topics and to create theories that subordinate the nature of any phenomenon under the explanatory framework of the disciplinary way of theorizing. To put it in other words: Transforming the different aspects of social life into a way of theorizing about things bends the phenomena of theorizing under the preoccupied views of the pre-assumptions manifested in the categories and theories constituting the particular ways of the distinctive ways of disciplinary thinking.

As the existence of different aspects of human life does not justify making them a theoretical perspective constituting the particular way of the disciplines to scrutinize the social, the existence of these different elements of any social, does also not justify the disaggregation of disciplinary knowledge.[63] Unlike disciplinary think-

---

[63] Since the emergence of disciplinary thinking as the cognitive architecture of social thought under the social sciences, disciplinary thinking discusses how

ing in the social sciences wants to believe, the disaggregation of disciplinary knowledge into distinctive disciplines is not due to the distinction of human live into economic, social, political or psychological aspects. Disaggregating theories about the social in to distinctive disciplines and their theories resulting from the presupposed disciplinary views on the social, that constitute the particular ways of theorizing of the different disciplines, the disaggregation of these theories is not the result of the different natural aspects of human live, but the result of a very *made* nature of human, made by nation state constructed societies.

Constructing the necessity of a re-production of human life, their material basis of livings, as conflicting and exclusive interests, the communal bodies they create as an intervention against their life priorities that regulates and restricts individuals for the sake of the ruling interests of their community, communities ruling their members against their own interests with nothing less but—a monopole of—power, is anything else but describing the multifaceted nature of humans nature, as the natural sciences do with the truly multi-faceted nature; and from there to argue for the need of a multiplicity of theoretical approaches representing those human instances into which they divide the human, argues not with the nature of humans, but obviously with the peculiar nature of a society that forces humans with its power instance towards the separation of such distinctive and though interrelated instances.

Presenting this most odd and peculiar ways of organizing a nation state society as the nature of humans and justifying the need for the diversity of exclusive theoretical frameworks for thinking, only documents the affirmatism of social sciences, which incorporate the particular format of how the society they reflect upon is reconstructed in their way of theorizing and thus reflect on these made humans through these very made society model constructs.

---

to overcome disciplinary thinking. The debates about inter-disciplinarity ever accompanying the insistence on disciplinary thinking, somehow confess that the disaggregation of social thought into disciplines violates that real social live knows the distinction in different aspects, but not the disaggregation of them into disaggregated social thought. It is telling that the debate about inter-disciplinarity always occurs, when social sciences reflect upon the problems to intervene with the disaggregated theories they create into social live that does not correspond with disciplinary thinking founding the categories through which disciplinary thinking creates its theories.

Chapter B: Categorical essentials of disciplinary thinking

That such affirmative thinking, founding social thought in thinking via disciplinary thinking on the assumption that justifies a particular society model as the foundation for a particular way of theorizing, inevitably results in false, that is in pretentious theories about the social, this society model practices as thinking about itself, is already obvious from the architecture of disciplinary thinking.

The cognitive basis for a plurality and the disaggregation of social thought in disciplinary thinking in the social sciences about the nation state social is to re-interpret the practiced abstraction of a society consisting only in this practically executed abstractions that transforms them all by disregarding their not only most different and opposing and exclusive economic and private life into equal citizens the nation states, an equality the political power sets into force, and to re-interpret this abstraction as the ideal of a community of humans and as the nature of human' race is what constitutes the cognitive basis of disciplinary thinking and its ever critical thinking. It is this secularized, methodological racism thinking the nature of nation state made subjects, their separation in economic, political and private subjects as the nature of humans that constitutes the plurality of the social sciences as the plurality of separated disciplinary social thought.

It is this peculiar construct of humans only the nation state made society model practices, abstractions executed at the humans, executed with the violence of the political power of nation states disciplinary thinking presents as equal() members of an idealized community that constitutes disciplinary thinking as thinking about a multiplicity of the different disaggregated aspects of human nature, the "roles", disaggregated into the multiple functions they are attributed as citizens and forced to practice as their life frames, it is these real state social constructs, that gives birth to disciplinary thinking, that is, both to the reflexive architecture of disciplines as to its categorical basis.

Not coincidentally the founders of the "modern" social sciences and of disciplinary thinking, as the later gurus of any epistemological leadership accompanying social sciences through their history are mostly sociologists. The abstraction of a society, an entity of equal-ized subjects is the violent creature of transforming all humans of the nation state society into equal citizens, subordinating them all under the political power of the laws of nation states via disregarding their very substantial economic differences and opposing life agendas, all the equalizing lawful definitions of citizenship any citizen very well knows, as does sociology.

Nevertheless, this false abstraction, does not only constitute sociological thinking, the creation of a community of equals that denies all their differences and their exclusive life agendas, that denies all conflicting interests, their power relations and alike sociological thinking then after abstracting from them re-introduces as if these subjects and these life agendas were not the construct of this very society model, but only were the conditions under which the naturalized humans struggle to arrive at their ideal of a community of humans and at humans which therefore need the supervision and advice of sociological thinking—as of the other disciplines. This false transformation of a society consisting of competing subjects with opposing interests into the ideal of a community gives birth to re-define macro-economics and political theory as theories looking at the social from the point of view of the very made societal subjects created by the nation states.

Thanks to the creation of the sociological construct of a community, thinking about the political power that forces the competing subjects under the rules of nation states, has been transferred with this abstraction of a sociologically defined community into the sphere of political sciences, which from there on discuss the political power as the technics of serving to solve the citizens life challenges, they only have because they are forced to perform their lives under the conditions of the competitive society and their opposing interests.

Economic theory, thanks to the construct of the sociological community and the implied shift of all power affairs towards political theory, reveals its topic as serving the economic interests of the communards, just as if the whole economic life was not the product of the political power, that forces its citizens to strive for their life aims via the economy the very political power established and supervises as the exclusive economic living condition, economic living conditions, which for most citizens turn out to become rather their life aim but their living conditions.

Thus, before any social thought is created by any of these social sciences disciplines social science thinking starts from the most fundamental social constructs of the nation state society, presenting their division in social, economic and political lives as the nature of humans, that—strangely enough—does not work very naturally, but needs to be set in in force and supervised by the ever requested power of nation states. For this, explaining why only serving the nature of human, though creates so many turbulences, a third social science has found its mission, psychological thinking, interpreting all problems on earth with—the nature of humans, which—not surprisingly—a thinking which re-detects all the features of creatures of the nation states society, the citoyen, as challenges of the natural nature of mankind.

It is these dis-aggregations of human life, only the nation state societies sets into force by its political power, it is this political power, which defines humans as equal citizen abstracting from their very different real means to perform their lives, that forces these people into an competition about gaining their living conditions in an economic battle against each other, in which the others loss is the others gain, a battle the majority of people can only lose thanks to their lacking means, the very political power knows very well while abstracting, i.e. ignoring and anticipating the inevitable outcomes of this battle, the exclusion from the wealth people produce, in the social welfare departments, which generously offer the continuous reproduction of the losers living conditions, perpetuating their secure loss as helping them to carry on with these cynical life circles. These are the abstractions, the disaggregations of the real life, the political power of the nation state societies puts on the stage of life, disciplinary thinking does not want to scrutinizes, but takes as the positive starting points, the "real facts" (Comte), constituting social thought thinking about the social through their disciplinary abstractions.

Instead scrutinizing the systemic objectives, this multiplicity of the nation state made humans obey and how their decomposition into a duplication of abstracted, that is separated, subjects, the political citoyen and the private, intervenes into their lives, disciplinary thinking elaborates the decomposition of the nation state human towards a multiplicity of creatures as their approach to theorize: Disciplinary thinking is populated by a multiplicity of subjects, cognitively reproducing the practical separations the nation state society executes with its political power, that defines who people are: Children, students, pensionists, employees, tax payers, employers, politicians, families, in short, all these creatures of the nation state society, disciplinary thinking does not want to scrutinize, but takes them as subjects founding their critical thinking about their "roles" for the society and the polis and by doing this, constitute the categorical foundation of thinking through the individual disciplines explanatory models, through which they approach thinking about the social in the social sciences.

From here on disciplinary thinking is populated with humans who seem to have a multiple disaggregated life: In this, thinking in the disciplinary disaggregated life scenarios, economic thinking confronts us with a homo economicus, whose political life as a citizens is elegantly excluded from any economic reflections, though it is only the power of law that forces him into earning the means for his life in a re-production system that is politically well organized on some carefully differentiated and lawfully regulated markets; on the other side of the disciplinary disaggregated human, there is a homo politicus, who appears in sociology, and in political theory as if his economic life was a mere condition for his freedom as a private creature—this, his economic life, was delegated to the other discipline, creating the topics for economics.

It is this, the disaggregation of humans into economic creatures here and social creatures there, all unified in a society of equal property owners, abstracted from the fact that most of them do not own any property, that gives birth to the social science disciplines, their categorical apparatus and their theories, that provide the cognitive basis for social science theorizing via disciplinary thinking, the pluralism of social thought in social sciences and the basis for thinking in social sciences about and through these nation state constructs.

Essential concepts founding theorizing in the classical social science disciplines

Not only the architecture with a plurality of social sciences views dividing social thought into disciplinary thinking reproduces the multiplicity of the practical perspectives the nation state subjects have on the social. The categorical apparatus through which each discipline thinks about the social consists of interpretations of the social through the idealized lenses of idealized nation states. The categories founding theorizing in the individual disciplines are variations to unfold theories about the social through a perspective only a nation state view on the social has.[64]

Anthropology—Regimen as the demand of man's nature

Anthropology is the only discipline among the social sciences that reflects about the social beyond the nation state socials—one could at least conclude from the fact that Anthropologist do find their theories beyond their nation state societies—mostly their imperial nation states—since Anthropological thinkers mainly come from imperial countries and their thinking gains its knowledge via research about "native" people, mostly and preferably in the most remote countries or areas of the world, untouched, as much as possible from the world of nation states, untouched by the society model of the imperial world.

For doing this, investigating the life of "native" people, anthropological thinking has been—rightly—accused of creating racist thought discrediting the "natives" as uncivilized and alike. This is no doubt the case, and if one reads the justification of missions of famous studies of some anthropologists, such as Malinowski or later Benedict[65], one may rightly wonder, if theories which explicitly

---

[64] It should be stressed that the following critique of the categorical essentials on which theorizing in the classical social sciences is founded, is neither any elaborated critique of those individual disciplinary theories, nor does it substitute the need to trace the thought of any individual theory. The following discussion of the categorical foundations of the individual disciplines only tries to demonstrate, why and how social science theorizing incorporates in its categorical essentials, on which thinking in these disciplines is founded, a way of theorizing that considers the secluded nation state social in a world of nation state socials not only as the unit of analysis but also as the theoretical perspective from which they construct their theoretical apparatus of categories through which they reflect on the nation state socials, that this is thinking that thinks about the social essentially through the idealized social constructs of a nation state social and that, thus, not only consequently off-thinks the world beyond and ends up in global social thought as imperial thinking, thinking through the perspective of a global power on the world's social.

[65] "The Japanese were the most alien enemy the United States had ever fought in an all-out struggle....It made the war in the Pacific more than a series of landings on island beaches. It made it a major problem in the nature of the enemy. We had to understand their nature in order to cope with it." Benedict, Ruth, (2005) *Chrysanthemum and the Sword, Patterns of Japanese Culture*, First Marine book edition, p 1

were created for assisting imperial powers to carry out wars, as not only Benedict frankly motivates her theories about Japan, deserve to be treated as scientific theories and to be cross-checked what these thinkers found in their studies. However, as much as one may be tempted to reject and not to critique these openly racist and militaristic theories and, as it is normally done, rather pointing on their racist and militaristic missions, ignoring their theories pays the price, that these theories until today, despite of their critiqued *intentions*, are quoted as providing insights, such as the study of Benedict, on the Japanese culture as about other people who are subjects of such theories. If at all, these theories are rejected as in the case of Benedict's theories, by many Japanese academics who, however, do not by critique these theories, but rather express a lack of understanding of who Japanese really are and, by doing this, these criticists share the approach and the topic of such studies, sharing the essential assumption of such anthropological studies, that there is such a national nature of humans, they later, accommodated to the categories introduced by cultural theories, call a "Japanese culture".

I hold that anthropology not only occasionally creates racist theories about mainly native people they investigate, but that the research question constituting anthropology as a discipline is a racist idea from the very beginning and that the racism of anthropology is the meta version, quasi a methodological racism, of the very same racism that is in no sense different from the racism in the presuppositions, the images of humans, constituting the other social science disciplines, that often accuse anthropology of creating racist theories.

Anthropological thinking is about a most fundamental question, as a fundamentally false question: Anthropology raises the question, what is the "exact image of man"[66], more precisely one should say, once anthropology shifted from the most obvious racism of social biological thinking towards social science theorizing, anthropological thinking asks—What is the social nature of man—and this is a false question. It is false, because it aims at finding behind all the social constructs *humans make*, human social constructs which represent a *genuine human nature*, constructs of human life that humans make, but constructs that coincide with any human natu-

---

[66] C. Geertz, *The Interpretation of Cultures*, Basic Books New York, 1973, p 44

ral constructs, thus; it is the contradiction of a natural necessity for any essentials in what humans create as their various ways of living. They seek for social constructs humans not only want and make, but—somehow—for what they must want and make, since it is what they naturally want and make, a life project that coincides with the human nature. They aim at finding, what all the variations of human life, created by humans have in common, that is to say, what the natural human nature is, beyond all the concepts of social humans invented and made throughout history.

Finding human constructs of the social humans somehow must want due to their social nature, a concept of the human social, they have "under their skin", is not only an odd paradox, but results in a methodological crux, since it aims a finding beyond the articulation and practiced models of human life a meta genuine human life model. Finding the essentials of a meta life project "under the skin", as anthropologists circumscribe their reflexive paradox of searching for what "man" *is*, is a question that abstracts from all particular real life projects of man so that nothing remains but the empty abstraction "man", empty, meaningless essentials of sociality, which are though supposed to craft all the various life models, and, inevitably, this is a question that concludes with the most shallow and tautological definitions of "man", definitions anthropologists like to then re-find in studying the human nature, preferably in the nature of humans, who are not much emancipated from their nature, the "uncivilized" natives, to prove that they are human's nature.

The way to handle this self-made methodological problem in anthropological thinking is as odd as the problem and demonstrates the essential fault of anthropological thinking: Anthropological thinking looks at manifestations of the natural nature of humans, that is, at manifestations anthropological thinking wants to detect, not as something human make, but as what humans quasi naturally must do. For this exact reason, anthropologists try to find the "exact image of man" underlying all constructs of human life, among those humans who are the least spoilt by what humans make, not the "civilized" humans and their life projects, but those, whom they believe to be still somewhat closer to human nature. This is why anthropologists find their human life essentials by studying the life of "natives", unspoiled by all the civilization projects, but though crafting them all.

Anthropology, historically a younger social science discipline, starts its thought about "man" from a critique of disciplinary thinking, since anthropological thinking does not think about any aspect of human life, as the other disciplines are supposed to do, but about "man" as a "unitary system", man as a whole and man as such:

> "...we need to replace the "stratigraphic" conceptions of the relations between the various aspects of human existence with a synthetic one; that is, one in which biological, psychological, sociological, and cultural factors can be treated as variables within unitary systems of analysis. The establishment of a common language in the social sciences is not a matter of mere coordination of terminologies or, worse yet, of coining artificial new ones; nor is it a matter of imposing a single set of categories upon the area as a whole. It is a matter of integrating different types of theories and concepts in such a way that one can formulate meaningful propositions embodying findings now sequestered in separate fields of studies."[67]

Anthropological thinking dislikes that the social science disciplines lack a synthesis of human existence, dividing the human into "sequestered studies" focused on "biological, psychological, sociological, and cultural factors." Enlightened anthropological thinking during colonialism, starting from biological thinking, suggests a mission of anthropological thinking that embraces the "sequestered studies" of the other social science disciplines and knows how to....

> ".... launch such an integration from the anthropological side and to reach, hereby, a more exact image of man"....[68]

To achieve a more "exact image of man" than the "sequestered studies" of the other social science disciplines anthropological thinking suggests to use a twofold new interpretation of culture:

> "The first... is that culture is best seen .... as a set of control mechanisms .... for the governing of behavior. The second idea is that man is precisely the animal most desperately dependent upon such extragenetic, outside-the-skin control mechanisms,..., for ordering his behavior."[69]

---

[67] Ibid.... p 44
[68] Ibid.... p 44
[69] Ibid.... p 44

Chapter B: Categorical essentials of disciplinary thinking

Anthropological thinking is unbeatable by any of those merely "sequestered studies" of the other disciplines re-defining culture as a set of exactly the very "control mechanisms" the animal man is desperately dependent on "for ordering his behavior". What a wonderful coincidence that the "plans, recipes, rules, instructions" the famous anthropologist Geertz defines as culture, match what he defines as the desire of man for "ordering his behavior", a coincidence, not that difficult to detect as the anthropological version of the one and only argument of all other social sciences, translating the relation of the ordering power of nation states and their subordinated citizens into a service for "man ordering his behavior", those disordered 'man' who, thanks to the creativity of anthropological thinking, created all those ordering "plans, recipes, rules, instructions" they are so desperately dependent on to order their natural disorder.

As already demonstrated along the notion 'globalization' the art of social science theorizing consists of the genius ability to disguise most obvious contractions under the vagueness of phrases, which disguise any subjects, any objects and any actions. Why and how does man, whose nature is to be disordered, detect his desire not only for "ordering mechanisms", why does 'man' not create the order he needs, but creates a desire for being subjugated under such "ordering mechanisms"; And: Who are those subjects, who, despite of their disordered nature, are not only able not only to create but to provide the naturally disordered 'man' with such not coincidentally subjectless "mechanism"; who keeps these "mechanisms" working, and so on and on. All this, all those absurdities a real master in social science thinking knows to present as sound insights with the precisely therefore invented vagueness of the categories he uses. Calling the "ordering mechanism" 'political power' is beyond anthropological thinking, since anthropology does not think about such ordinary things, but thinks about such "thick" issues like "man".

It is this one and only message that anthropological thinking varies throughout all its non-sequestered studies, which jungle they ever invade to study "man" and to find again and again these insights about "man". What do they find? All kinds of "natives" are, whatever they do, engaged in any adjustments between the "outside-the-skin control mechanisms" and their under-skin desperate desire for "ordering their behavior".

While the founder of anthropological thinking, yet unaffected by notions such as culture and behavior, detected among some natives in New Guinea Islands in their

> "...belief in magic ... one of the main psychological forces which allow for organization and systematization of economic effort"[70]

the more contemporary anthropologist, Geertz, detects in his studies about cockfights in a remote village on Bali after more than 30 pages with "thick description" exactly the same, now phrased in the updated category fashions of the contemporary social sciences.

Geertz studies people on Bali and finds things like this in his "thick" observations:

> "There is a fixed and known odds paradigm which runs in a continuous series from ten-to-nine at the short end to two-to-one at the long: 10-9, 9-8, 8-7, 7-6, 6-5, 5-4, 4-3, 3-2, 2-1. "[71]

One can understand that such observations about what the "exact man" genuinely is, can only be found among "natives" on Bali and one can almost sense why no health risk can hinder the missionary project of anthropological thinkers, to find out what the "exact man" is:

> "Early in April of 1958, my wife and I arrived, malarial and diffident, in a Balinese village we intended, as anthropologists, to study."[72]

And, what did the anthropological couple find, "as anthropologists", under the skin of the Balinese people while observing cockfights:

> "...it provides a metasocial commentary upon the whole matter of assorting human beings into fixed hierarchical ranks and then organizing the major part of collective existence around that assortment."[73]

Bali people do not only do exactly the same as what Malinowsky's "savage people" in New Guinea do, but they do what all those people do, who are forever and everywhere "deeply" studied by anthropologists, may this be while constructing canoes, celebrating

---

[70] Malinowsky, Bronislaw, *(1966) Argonauts of the Western Pacific,* Routledge London, p xxi
[71] C. Geertz, *The interpretation...* p 426
[72] Ibid... p 412
[73] Ibid... p 448

funerals, enjoying cockfighting or practicing ceremonial myths or religion, the latter one a preferable topic of anthropological studies to learn about the enlightened "man": They all struggle, thanks to their definition of what "man" always and everywhere does, to refind what anthropological thinking is convinced "man" is, whatever "man" does and wherever he does he does the same, in any jungle or in Paris or elsewhere.

In Geertz's definition of culture, a mélange of the latest fashions of a sociologically phrased version of culture, "man" exactly does whatever and wherever "man" does anything, "man" ever and anywhere is assorting "man", assorting his "collective existence" to "the set of control mechanisms", or to "fixed hierarchical ranks" and—phrased in the fashion of behaviorist psychology—"assorting" "man", "ordering his behavior".

And so on and on, one anthropological study after another, always finds the same about "man", varied just in phrasing the same findings with different categories, adjusted to those science fashions, which are setting the knowledge standards and providing the categories for the social sciences discourses. Thus, Geertz rephrases what Malinowsky already "found" in the jungle he visited, accommodated to the categories of the latest social science fashions.

As we shall see when discussing the "sequestered studies" of the other disciplines anthropological thinking may believe it found a meta approach to present the naturalization of nation state subjects as the nature of "man", not only concluded from any "aspects of humans existence", but a "synthetic one"; that is, one in which "biological, psychological, sociological, and cultural factors can be treated as variables within unitary systems of analysis." [74] However, anthropology only repeats the false thought of the other disciplines, trying to present the nation state project and its creatures as a godlike present for humans, incorporated in his nature, a gift for the nation state inhabitants to cope with their inside-skin behavioral disorder, a disorder anthropologists par tout do not want to find in the nature of the competitive society that the nation state makes and "controls" and whose creature they observe, but in the nature of the "exact man".

---

[74] Ibid... p 44

To put in other words, what anthropological thinking finds in its studies about "natives" beyond the established imperial nation states, may this be on Bali, the New Guinea Island or elsewhere, is the finding that the natural desire for what anthropological theorizing beforehand defined as "culture", a desperate desire not for the real society model, the imperial nation states practice and spread across the world as the world's way of conceptualizing human life, but for a society model, that interprets the project of nation states ruling citizens life towards serving the nation and its economy, as assisting humans to deal with ordering its natural behavioral problems, problems humans are attached by Anthropologists as their "exact" nature, a human who ever seeks for nothing but ordering his naturally dis-ordered behavior, and who coincidentally finds this in the "outside-the-skin control mechanisms".

Why this naturally dis-ordered human feels all against his disorganized nature the desperate desire for ordering his behavior and how this human, carrying disorganization under his skin manages to create this "outside-the-skin control mechanisms", are questions, anthropologists do not need to answer, since they found these contradictions, not in any studies about "natives", but simply borrowed them from another social science discipline, from a contemporary fashion in psychological thinking, behaviorism and from cultural theories. The "outside-the-skin control mechanisms" such as "plans, recipes, rules, instructions" is a way to re-phrase the ordering power of nation states as a cultural achievement, an offer for the ruled, is only difficult to read for thinkers as an euphemistic, idealized image of nation state societies for those professional thinkers, who elsewhere preferably talk about culture in nationally constructed entities, but do not even want to distinguish between "plans" and "instructions", though at the latest thinking "instructions" as a "desperate" dependence upon...extragenetic, outside-the-skin control mechanisms" must sound a bit irritating—unless anthropologists uncritically borrowed some insights from behavioral psychologists. No, cultural anthropologist are blind to such obviously distorted notions, since they are determined to ennoble the nation state societies, just as if they were the same as a football club or any other "assorting" "mechanism", into a cultural achievement for "man", a service for asserting "man" to the otherwise disordered "collectives".

124 Chapter B: Categorical essentials of disciplinary thinking

Just like all other social science disciplines, the thinkers thinking about facts, do of course not just ennoble nation states or any particular nation state. Like all other social sciences, anthropological thinking does not live in this world, but in the world of the most fundamental ideas, dreams, imaginations about "man" and therefore rather celebrates not any profane nation state as the realization of the dream of man, but resonates more about ideas lying behind such vague things like organizing "mechanism", to tell us what 'man" really after all is.

About "man" they found

> "We are in sum, incomplete or unfinished animals who complete or finish ourselves through culture. ... Beavers build dams, birds build nests, bees locate food, baboons organize social groups, ... But men build dams or shelter, locate food, organize their social groups....under the guidance of instructions encoded in flow charts and blueprints, hunting lore, moral systems and aesthetic judgments. Conceptual structures molding formless talents."[75]

Needless to say, that the same thinker knows that "conceptual structures" are "manufactured", made by the "formless talents". However, perfect in form thinkers, looking behind man, prefer to stick to their contraction of a formless talent that is molded by "cultural artefacts", leaving the question of who manufactured them within the darkness of a mysterious fairy tales about "man", which allows them to present their admiration for the godlike artefacts as the heaven for the "unfinished animals".

Then, answering the question, regarding which part humans miss as incomplete animals, anybody who remembers the fairy tales from his grandma, remembers fairy tales about the "bees" and the complete beehive, possibly thanks to the contributions from German anthropologists, called "Bienen*staat*", humans are incomplete because they lack what the bees own by their very nature,— exactly—at nation state, which incomplete humans can though construct, thanks to the "guidance of instructions encoded in flow charts and blueprints" and thus also arrive at the organizing mechanisms bees are given by their already completed nature.

Any very complete animals must have been so wise as to programme "instructions encoded in flow charts and blueprints" allowing to achieve thanks to their adaptation talents what bees have

---

[75] Ibid... p 50

thanks to their nature. And so on and on. One should not forget, this is not grandma narrating, but one of the world's wide most distinguished anthropologists. Who wants to believe that this must be any dark moment of theorizing should read other Anthropologists—and will find everywhere exactly this Christian image about the nature of humans, which is the foundation of anthropological thinking, since unlike the animals and humans they think about, anthropologists are another exception from both.

Anthropologists, just as the programmers writing the "instructions encoded in flow charts" obviously must be another exception to the "unfinished animals" and though they obviously know very well know who the real subjects writing "instructions" are, elsewhere called laws, they make them up as a choice for any man, instructions between the incomplete human may choose, just like birds choose between nests:

> "...we do not attempt to explain on a genetic basis why some men put their trust in centralized planning and other in the free market, though it might be an amusing exercise."[76]

It is rather an amusing exercise not being explained "on a genetic basis" by those naïve grandmothers disguised as anthropologists, which genes make up the "formless talents" to "put their trust" in communism, others in capitalism. It is however worth noticing that as much as the world of anthropological thinking lives in the realm of fairy tales like raisonnements, that musing about opposing society system anthropological thinking very well knows to distinguish, that their alternative "artefacts" are in the end various concepts of nation states.

However, anthropologists are anthropologists and live in the clouds of their mystical ideals where they muse about man and would never come down to earth and to think about such profane things like nation state, society systems and other real things. They do not have to celebrate the real nation state, or even the concept of nation state, the unfinished animals must appreciate as the "control mechanism" they must trust in, since they have already ennobled *any* "cultural artefact" into a godlike present for the "unfinished animals".

---

[76] Ibid

Not even in their studies, in which anthropologists describe in the most detailed ways their observed "native" people ever organizing and ordering, even when they practice their magic rituals, anthropologists find nothing else but an early version of "Primitive Law and Order"[77], ever driven by nothing else but their one and only horror vison of non-ruled people, desiring above all rules.

Man's nature is to be disordered and therefore his nature is to need order: That anthropologists have found the same as other social sciences, now as "one in which biological, psychological, sociological, and cultural factors can be treated as variables within unitary systems of analysis", is easy to see through as the life lie of anthropology struggling with its existence as a discipline, once the concept of nation state has been globally completed, also among their preferable witnesses for what "man" is in all the remote, natural locations remaining on the globe.

No doubt, anthropological theories create racist theories, about the native people they enjoy to observe in order to re-find their ex ante definitions as their theories about man, however, the racist theories of anthropological thinking do not only apply to the native people they investigate, their racist theories are much more fundamental and do not only apply to any human, when they claim to scientifically determine what is human and what not, but are theories that can only construct their racist thought, not only with their ignorance about the native people they study, but with false categories they entirely borrow from other disciplinary theories and apply them to their meta theories about "man" in their anthropological studies.

From anthropological thinking to cultural theories— nation states as cultural artefacts completing man's unfinished nature

Culture is the major category more contemporary Anthropological thinking embraces, an embracement that finally dissolves anthropology as a distinctive discipline replaced by cultural theories.

As any social science theorizing, anthropological thinking is accommodative thinking. With the establishment of the post World-War II US world model which transformed the colonial world into

---

[77] B. Malisnowsky, Argonauts... p xi

a world of nation states, anthropological thinking declined. Due to the essential idea of anthropological thinking, to do research about the nature of man, they felt natives represent man's unspoiled nature, thus the natives were considered as their resource for finding answers on their racist question, who is man. For this they borrowed theories from other social science disciplines and created from their some patchwork-like theoretical frameworks, anthropologists like to present as "unitary systems of analysis", and take this on their trips to the decreasing natives on the world to find that "man" is what anthropologists had defined. The ironic tragedy of anthropology is that there theories have been made true by World War II making the world into one world where all humans are served with what anthropologists like Geertz likes to phrase in his kind of psycho-techno language as the "extragenetic, outside-the-skin control mechanisms". The world was made a world of nation states and the humans, living in this world since then were all provided with "extragenetic, outside-the-skin control mechanisms", with the effect that the natives increasingly disappeared as the object, spring and proof of anthropological thinking.

> "In sum, in the first decades of the 20th century, with different nationalisms and colonialisms at work, natives were mostly viewed through modern eyes as peoples who needed to be known in order to propitiate their integration to nation-states or empires: ...Anthropology's real triumphant and booming period started after the Second World War. In part it coincided with the modernizing drive of the time that called for educated masses that had greater access to a rapidly expanding university system in many countries. But it also coincided with a renewed demand for 'scientific' knowledge about strange and exotic natives for the sake of 'development' needs worldwide. Increasingly, natives ceased to be colonial subjects of Western empires and instead became citizens of 'underdeveloped' nation-states."[78]

While anthropology thus gradually vanishes, its essential question survived and the question, what is the nature of man, was adjusted to the practiced globalization of the identity of man as a citizen of nation states. Theories seeking answers about the question who man is, before investigated among the colonialized socials, were replaced by the question who is the *national man*. Thus, the racist question of anthropology, methodologically applied to the not yet nation state citizens to prove that nation state citizens coincide

---

[78] Ribeiro, Gustavo, *Critique of Anthropology*, http://coa.sagepub.com/content/26/4/363.abstract, p 367/8

with the nature of man, was now increasingly replaced by the enlightened racism of cultural theories, generalizing the racist thinking about the exotic otherness to the whole world of now everywhere nationally constructed socials beyond and within any nationally united social. Who are the Japanese, the Germans, the Indonesians, is the modernized issue of the anthropological question who is man, now called cultural theory, applied to all national man.

With the universalization of nation state societies all "men" are nation state "man", the world consists of nationally constructed "man", which inspired cultural theories to update the question of anthropology and to think that it must be the concern of man that man does not know anymore who all those nation state constructed men are. Hence, the more ontological racist question of anthropological thinking was updated into a more practical racist view on understanding the many national natures of all the nationally constructed creatures now populating the whole world.

Indeed, nation states have their own traditions and interpretations, though all of the *same* nation state concept and it can be helpful to know them if man meets man. Cultural theories, not interested in knowing what a nation state human is, since anthropology had found out that nation state humans are coinciding with the nature of humans, serving them with control mechanisms and such things, man's nature needs, cultural theories detects in this a challenge for a new version of the old question of anthropology, the more recent version of anthropology, such like Geertz's, had already created from a mélange of fashionable sociological and psychological categories such ideas like "the desperate need of set of control mechanisms .... for the governing of behavior". With this updated version of anthropology, anthropology was transformed into cultural theories, which carries the anthropological ontological issues now back from the outbacks into the more practical views of more pragmatic thinkers assisting "man" in the global arena of internationally acting citizens.

Cultural theories re-interpret the racist question of anthropology, who man is, into the question, which national man is who, a new challenge cultural theories detects to govern the behavior of global man and detect a need for cultural theories to create an updated version of no longer a control-mechanism but of an orientation-mechanism, allowing all the nation state man to understand who nation state men are.

Assuming that men, for cultural theories, is above all and essentially a nation state creature and as such equipped with a multiplicity of different national natures, cultural theory therefore creates a theory that aims at finding out first how national humans represent a particular nation state, national characters of nationally constructed men, and, second, *how to escape from and to foster* this nature as a nationally constructed creature among others of the same kind. Cultural theories therefore creates nationally designed stereotypes about any ever nationally constructed social entities, politically or ethnically constructed "nostri", created and fostered by the tolerant knowledge of cultural theories, tolerating, relativating and fostering those stereotypes only cultural theories construct about these men, only cultural theories interpret as their nature.

Culturally defined creatures, generating with their national nature distinctive human communities, defined along naturalized national natures, populate since then theorizing about the world's social as the subjects of the enlightened social science thinking in cultural theories across all disciplines. [79]

Subsuming humans into any cultural entities, such as the European science view gathering "Asians" into one cultural entity, politically constructed entities presenting the view of imperial thinking as the nature of these inhabitants, only an imperial categorization of the inhabitants of other national social can construct, discloses the ignorance such thinkers have about their own entities, they can only invent thanks to their very ignorance and prejudices about them.

However, ignorance and prejudices are not only the cognitive resource of cultural theorizing, but are also its major result: Subsuming their invented humans under the prejudicial stereotypes that cultural theories create, does not even try to hide the moral value judgements as the only intellectual resource that creates such stereotypes. Stereotypes such as "collectivism", any cultural theory appreciates as a valid theory about "the Asians", bring the former exquisite racism of anthropological thinking to a perfectionism, ever measuring not only the natives, but all humans across the world

---

[79] Accusing these social sciences of their "orientalism" whenever they paid attention to the "otherness" was a well-received book—however, only in the "shocked" "Western" social sciences; an opportunity for the humanistic social sciences to radically distance themselves from such kind of "excotism"—and to continue with the same.

as deviating from those human ideals, cultural theories construct from the shallow ideals of the imperial world, such as the idealization of the imperial socials, as "individualist". As if both collectivists and individualist were not the two sides of the same morally constructed dichotomy, coining individualism as positive and collectivism, thanks to its discrediting meaning in the West about the East, negative, is a dichotomy that only exists in the fantasy of both determined *racist and tolerant* thinkers, translating the *tensions within the* dualism, not cultural theories but sociological thinking invented, into alternative human value judgements only cultural theorists attach to people, dependent on the value preferences, negative or positive, they want to attach to them.

As if these two features, "individualist" *and* "collectivist" were not *the* essentials of the nation state constructs, the citizen, requested to combine individual and communal features, cultural thinking takes this tension, invented by sociology as the obscure conflicting nature of humans, attributes to them the very value judgments they find in political prejudices—(who does not know this: Collectivism are the underdogs or ants (Asians) of communism)—a negative connotation to those people they want to discredit and a positive connotation to those they obviously prioritize, the individualist (individualism are the heroes of freedom), thus quite frankly distinguished along this value judgement, that allows to discredit the "Asians", as the others, as opposed to the positive value of individualism, representing the human in the imperial countries, both easily taken from the public discourse prejudices in the imperial Western world.[80]

As if they wanted to prove that the regime of social science thinking in cultural categories is truly global thinking, scholars from Asia do not critique the stereotypes, naturalizing the essential of nation state humans into the distinctive cultural features of the nature of humans, but they accommodate their response to the racism of cultural theory thinking and do not reject the racist method

---

[80] As if "non-essentialist" cultural theories were created to prove the above mentioned conclusion about anthropological thinking that racist thinking towards a social outside of the national social provides racist thinking within each national social, "non-essential" cultural theories critiques cultural theories thinking in national entities and makes racist thinking to the essential of cultural thinking about any ephemeral nostricism, within and beyond any nationally constructed "nostrum".

of cultural thinking, but reject the negative value judgement about "collectivism" and use the same value judgements with different connotations to counter-discredit the discreditors: Individualism represents the irresponsibility of humans towards the "community" is the harsh counter-racism of the discredited confirming the general racism of arguing in cultural theories categories.

Cultural theories thus provide, above all, unlike the ontological theories of anthropology more applied theories about "man" and create preoccupied social thought that preserves and modernizes the racist stereotypes of anthropology towards an updated version of what man is, updated for the global battle among all kind of national men, involved in all kind battles among national men.

The more the world is unified, just as if difference did not exist before, wherever cultural theories detect difference, mystified by ignoring their common essentials, without which difference would not be difference, any detected difference is ennobled by cultural theories towards a "cultural artefact" and thus serves the unified social world, forced under the very same objectives to appear as mystic monades of exclusive and particular identities, cultural theories firstly create, then help to decipher and advocate to generously tolerate.

Economic thinking in the social sciences— The bane of scarcity

Thinking about the economic elements of social life, the social sciences present the disciplinary view of economics as one responding to the economic nature of humans and yet reveal that their economic thought reproduces the views that nation state creatures, the subjects of the private property ownership have on a particular type of economy, an economy nation states impose on humans as the mode of re-production, distribution and consumption, not at all only as the economic *conditions* of their life aims, but as their life *essential* towards which they must subordinate their life agendas.

For social sciences theorizing about the economy, economic life is all about scarcity and scarcity is for economic thinking in the social sciences not thinking about the nature of a particular type of economy, but for economic thinking in social sciences thinking about scarcity is thinking about the nature of any economy.

"However, if we boil down all these definitions, we find one common theme: Economics is the study of how societies use scarce resources to produce valuable goods and services and distribute them among different individuals."[81]

It is not the case that economic thinking in the social science does not know that the particular society model of capitalism is only one type of economy out various ways to economically reproduce different types of "societies". However economic social science thinking mentions the plurality of different types of economy, only to state that they all share the same problem, the problem of the *"scarce resources"*.

While practically any economic activity in the capitalist economies obviously revolves around money, the conceptual essential of economic thinking in social science economic theorizing about an economy of a society in which all social objectives are ruled by and dependent on possessing money, for economic thinking is not at all money, but "scarcity". Scarcity is a quantitative relation between the "resources" the economic subjects create and possess, a never reachable amount of "resources", whatever efforts the economy might make. Within the concept of scarcity, there is not too little of any particular resource, but ever too little of any resources, not only disregarding what the productive usage of these resources is.

The concept of scarcity does not lack any amount of any particular resource, scarcity does not mean that there is not enough of this or that, but the concept of scarcity states that even the creation of an increasingly growing more of the outcomes of economic activities results in ever too little. How do they then measure an amount as ever being too little once it is counted as a quantity that is too little because it is not more than it is? Too little as compared to what? To more? To more, without knowing how much more, but in any case too little? What are economists measuring when they measure scarcity? What is the substance of the quantity economic social science thinking considers as ever being too little? As if economic life has not proceeded from collecting goods offered by nature towards producing goods, despite of the *production* of goods, for economic thinking about the economic aspects of human life any production results in the re-production of scarcity, not too little of any particu-

---

[81] Paul E. Samuelson and William D. Nordhaus, (2010) *Economics*, Tata McGraw Hill Edition, New York, p 4

lar good for any particular need, but ever too little as such. Ever too little as such, measured against what?

Like all social science thought, economic thinking about the economy presents the *homo economicus* as the nature of mankind and the economic nature of mankind is to ever count goods with a an ever beforehand fixed result to ever have too little—scarcity ever results in scarcity, despite of all economic actions, despite of an ever increasing production of goods and despite of all the economic expertise and insights ever economists ever create about the economy?

Scarcity, according to economic thinking in the social sciences, the essential of any economy, circumscribes that any quantifiable quantity is not enough, because it is not more than it should be. Scarcity is the—negatively phrased—ideal of an economy that aims at an ever growing growth of the therefore ever negatively counted resources, lacking resources economic thinking in the social sciences transforms into the dream, the wish that the created goods may be ever more than they are, an idea, founding thinking about the economy in economic theories that does not guide or assist any economic practices, but that justifies with this idealized objective, they present as the nature of any economy, an economy, in which the economic actors must make their economic activities a never reachable service for economic aims, motivated by the promise that it is the nature of any economy to never reach them.

Economic thinking in the social sciences therefore establishes a restless quantifying, a never satisfiable measurement of the relation of what the economy has and what it ideally should have but never gains, how much it ever creates, as the essential of economic theorizing.

Certainly, counting goods is an essential matter in the production and distribution of things a society consumes. Measuring the goods a society consumes and therefore needs to produce and to distribute is essential for economic thinking. However, the construct of social science thinking about the economy, saying that the result of any counting already exists before any counting, that is, that any amount is never enough, means that any economy ever recreates a scarcity no economic action can ever surmount, makes only sense in an economy in which not only the quality and use of the counted things are not the substance of counting, but in which a view on these things reigns this counting, for which all things are firstly the same and therefore any amount of these equalized "resources" call for a more. Only from the perspective of striving for abstract wealth as the aim of all economic actions, for money, considering all produced goods as the mere substance for an abstract *more, growing growth,* that only cares about the utility of things if they promise a more of abstract value, money, producing things ever creates the relation of scarcity, a more that is ever too little, since it is not more than it may be.

The essential of economic thinking in economic thinking, the concept of scarcity, is a theoretical ideal, constructed by economists who look at the economy through the eyes the practical view private property owners have on the economic re-production of human life, who are bound to strive for money as *the* aim of any economic actions: Only from the point of view of "growth", from an economy that measures its economic activities in growing quantities of abstract values, only from this idealized view of privates whose life depends on striving for money, there is never enough.

And not even this is fully correct: There is only one economic subject that aims at growth, an increasing amount of values. While the private property owners very well define if the growth of their invested money was a success or not, it is only the economic subject of the nation state that considers an ever increasing growth, not of any particular investment, but of the economy as a whole as a criterion to judge about the economy. Only the economic view of nation states on the economy considers any more as a growth that ever lacks an increasing rate of growth.

It is this speculative view through the practical perspective of the representative of the whole of the private property owners aiming at a growth of wealth measured in quantities of money that is never enough because it *should* be more that establishes this very idealist view on the economy as one view of social thought in the social sciences that separates thinking about the economy from the political and social elements of human life, constituting economics as one individual disciplinary perspective in the multiplicity of social science thinking.

Strictly speaking, not even the nation state shares this idealized view on economic growth as an ever repeated scarcity. This, the conceptualization of growth that is ever too little, is the invention only economic thinking in the social sciences creates that transforms the view of nation states on growing growth as measuring the success of the privates as a whole into a necessarily never fullfillable dream they though must strive for, an economic challenge of mankind, ever warning mankind to ever increase their efforts serving the demands of an ever growing growth, the bane of scarcity mankind can only ban by ever combating scarcity with an increasingly growing growth.

Once scarcity is established as the repeatedly recreated final reason for economic actions, economic thinking as the disciplinary view on the social in the social sciences on the economy can no longer be bothered about any phenomena this economy creates in its real life, namely those, where not an ontological, but a real scarcity can be observed. An economy that re-produces poverty in societies across the world and that creates and blasts values no society in history was able to produce and to destroy, an economy that transforms the world into a battle among national societies about the never solvable problem of "scarcity", for economic thinking in the social sciences all this is the result of their natural economic scarcity. Not only a world consisting of nationally constructed societies must fight against each other for their national growth and do this via treating the world's people as a scare resource ever exceeding the request for scarcity.[82]

Detecting that scarcity is the nature of any economy, the categorical foundation of economic theory is created as a distinctive social science discipline, founding the mission of theorizing about the economy as on an idealized economic agenda of nation states interpreted as a never ending battle against scarcity. Scarcity symbolizes an economic thinking as an ever critical warning towards economic practices to no violate the nature of economy, economic thinking appears as the raised finger symbolizing the moral rules economic activities must obey, the ever ex ante interpretation of the economy that is typical for economic theorizing, ever critically interpreting any economic problems, after they occurred, never

---

[82] The critical department of economic thinking, represented by such economists like Stieglitz argue that "rising income inequality is one of the main factors underlying the economic and financial crisis in the United States", is a variation of the very social science economic thinking presenting scarcity as a threat for mankind. It is the interpretation of people's income as—a factor of and for growth, a perspective that advocates economic policies to harmonize the economic outcomes for those who serve the growth and those who benefit from it—for the sake of combating the bane of scarcity.

saying how to match with what the nature of economy, they know so well, is and demands. Thinking about any economic phenomenon is established as thinking through the perspective of an idealized nation state rationale, interpreting an economic activity as coinciding or deviating from the aim of growing growth, the practical translation of the theoretical notion 'scarcity'. Hence, economic thinking from there on is about critically assisting idealized economic nation state rationales, critically observing economic policies if they obey the bane of scarcity, currently mainly by inventing econometric models that allow to calculate how to match the demands of scarcity, that is more concretely, to critically observe the relation of demand and supply on the various markets, those "markets" which are the economic means against scarcity and govern the economic life in a market economy.

Sociological thinking—The blessing of the "community"

Set aside what might be the reason for the coexistence of an uncountable growth of wealth and globally increasing poverty, the tautological explanation of economic thinking in the social sciences, that this is due to the nature of any economy ever re-producing less than more, might be understandable for nation state thinkers for whom the growth of growth is the ideal vision *and* the malediction of mankind. Social science theorizing seems to be theorizing of fundamentalist mystics passionately arguing against the threats of mankind's nature only they create and only they are coping with in their social thought.

While economists know to interpret the coexistence of poverty and wealth with a simple, 'there can never ever be enough', with scarcity as the nature of any economy, sociological thinking knows another natural aim of humans, which is that human nature is whatever they do to ever finally aim at being part of any social entity, an aim attached to humans that presents the nation state as this ever imperfectly realized social entity serving the demand of the very human nature to be social.

Society, community, group, individual, institutions, such things are the *métier* of sociological thinking.

"Neither the life of an individual nor the history of a society can be understood without understanding both." [83]

The understanding "of both", the *relation* of the individual and the society, is what sociological thinking coined with the categories like behavior, structure and the like. How individuals *relate* to the society and vice versa is the concern and the object of sociological thinking.

"Whether the point of interest is a great power state or a minor literary mood, a family, a prison, a creed—these are the kinds of questions the best sociological analysts have asked. They are the intellectual pivots of classic studies of man in society—and they are the questions inevitably raised by any mind possessing the sociological imagination." [84]

The society, the social structure, the relation between them and such things is, however, not only what sociological thinking is thinking *about*, the social is also the key to understanding the social, and, modest as sociological thinking is, without "sociological imaginations", no social can be understood—what raises the question, what is the social relationship between the sociological man in society that needs to be understood to gain the exquisite analytical spirit for "sociological imaginations", a spirit, as sociologists claim, without which no social thought can be thought?[85]

To anticipate the result, the answer on this question will also show, why it contradicts with the nature of sociological thinking to think about any social, including the world's social, other than through all the views of nation states entities without which sociological thinking could not think about any social.

Though sociological thinking is about how both, the society and the individuals, relate to each other, the dualism implies a twofold

---

[83] C. Wright Mills, (1959) *The Sociological Imagination*, Fortieth Anniversary Edition, Oxford University Press, New York, p 3. It is not difficult to sense from this statement, that it implies a distancing from psychological and historical thinking, who—according to sociological thinking—believe to understand both independently.
[84] Ibid, p 7
[85] It is this idea of the "social imagination" only sociologists have, which founds the scientific arrogance of sociology in front of all other social sciences disciplines and which results not only in the fact that sociology incorporates all other disciplines as a sub-section of sociology, but also in the pretentious misunderstanding that sociology and social sciences are—for sociologists—the same.

assumption, regarding the society and the individual: The first implication is that is not the individual that determines what the society is and aims at and that it is, secondly, ever the individual that adjusts its individual life plans to what the society determines. If the society, the community of individuals was the result of aims shared by the individuals constituting from their aims their society or community and all those other sociological entities gathering individuals, there was no need to relate the individuals or their interests to the society, since the society is the society thanks to their shared interests.

The humans in sociological thinking ever seek to adjust their biographies, to phrase it in sociological terms, to what the communities offer as a ruling structures. And it is the mission of sociological thinking to assist the subordinated to adjusting the life to the structuring services of all kind of communities and their structuring orders.

> "The problem of order, and thus the nature of the integration of stable systems of social interaction, that is of social structure, thus focuses on the integration of the motivation of actors with the normative cultural standards which integrate the action system, in our context interpersonally."[86]

The society, social structures and all the other sociological entities ever consist of a service, that is nothing but, integrating, structuring, organizing etc. the live of the individuals, assuming that individuals are no structuring entity in the relation of both, but subordinates who require to being the object of integration. The sociological constructs such as society, community, system or structure, all such concepts sociological thinking heavily argues about among the different sociology fractions, though uses them more or less synonymously, are all nation state entities, and all share for sociological thinking the same mission, that is to ever structure the relation of the society and the individual.

It is the weird concept of nation state to do nothing but providing order and, as a consequence, of all the nation state entities sociological thinking reflects about, sociological thinking calls communities, systems, structures, institutions, or also such things like families, the workplace, marriage, in short all the lawful nation state entities, for sociological thinking all essentially *are aiming at*

---

[86] T. Parsons, (1957) *The social system*, The Free Press, p 12

nothing else but providing order for the private lives of their citizens.

It is the essential fault of sociological thinking to construct political power, translated into sociological categories like, structure, order and the like not as imposing and executing the nation state objectives within the life of its citizens, but as a way to structure, to organize all these imagined most abstract and purposeless entities such as "structures" or the real community of citizens, as a political power that essentially does nothing else but organizing the otherwise non organized life of individuals. This essential mistake, considering nation states as a sort of steering committee on behalf and for the sake of an otherwise dis-ordered citizens, is constitutive for sociological thinking, always constructed with the same false conclusion: Imagine the very nation state creatures, the competing private property owners and their conflicting interests without the nation state, this would be chaos. Conclusion from off-thinking the nation state creatures without its creator, is to fortunately find that there is a subject caring about the relation between both, the nation state that helps to organize the life of the otherwise disorientated citizens—just as if these very humans and their opposing interests were not the very product of the nation state, not to mention the obscurantism making the nation state an institution of and for the disorientated citizens, who are though oriented enough to create this entity to orientate the disorientated.

> "...What ordinary men are directly aware of and what they try to do are bounded by the private orbits in which they live; their visions and their powers are limited to the close-up scenes of job, family, neighborhood; in other millieux they move vicariously and remain spectators: Seldom aware of the intricate connection between the patterns of their own lives and the course of the world history, ordinary men do not usually know what this connection means.... They do not possess the quality of mind to grasp the interplay of man and society ..... The sociological imagination enables us to grasp history and biography and the relation between the two within society. That is its task and its promise...."[87]

Sociological thinking aims at providing social thought for the "spectators" off all their communities, for people who have "limited power and visions". Sociological thinking never critiques the existence of "spectators", of underdogs, "bounded by the private orbit", but at serving the spectators with the limited visions and power

---

[87] C. Wright Mills (1959) *The sociological* ...., p 1–6

with sociological imaginations, such as the one about the sociological reason for the need of all the communities for these spectators. A typical example for advocating the spectators and their communities in current sociological discourse is to translate unemployment into social exclusion. For sociological thinking the problem for the subordinated is not their lacking income, but to be no longer part of all kind of communities, their "social exclusion". Just as if being unemployed is not an essential element of capitalist working life for people who cannot survive without a job, sociological imaginations interpret not being employed as a challenge for the integrating mission of the community as for the unemployed being excluded from communities, a sociologically imagined entity, into which the workforce of any company has been sociologically translated.

Classically, the idea to present the nation state as being challenged by any obscure non-sociological human nature that requires a structuring agency, the nation state that needs to "integrate" disintegrated humans assisted with the "task and promise" of sociological thinking enabling the spectators *"to grasp history and biography and the relation between the two within society"*:

> "... modern society is characterized by disintegration of community, the birth of individuals, individual's reintegration into nation-state,..... The traditional community is dissolved into individuals, and they must be reintegrated into the nation-state."[88]

As if the subject giving birth to the individual, sociological thinking is ever operating with but has no clue about, the competing private property owner, as if this individual, the competing private property owner, was not a nation state's construct of humans, the nation state ever appears as a deus ex machina ever requested to solving the challenge to reintegrate the otherwise disorientated citizens into the nation state, sociologically phrased, into communities..

Certainly, there are many life aims that require a community of humans to achieve them. However, the community humans strive for in sociological thinking is not only no community of people who share the same objectives and therefore jointly strive for their

---

[88] S.Yazawa, (2013),Civilization Encounter, Cultural Translation, and Social Reflexivity: A Note on the History of Sociology in Japan, In M.Kuhn, and K. Okamoto, (eds.) *Spatial Social Thought, Local Knowledge in Global Science Encounters,* Stuttgart: ibidem, p 133

aims; in sociological thinking being part of a community as such is *the* life aim of individuals, individuals who at the same time are disintegrated from their community. In other words, in sociological thinking humans are not a member of a society to reach any life aims they share, but being a society member for sociological thinking *is* their life aim that—strangely enough—ever conflicts both with the life aims of the other society members as with the aims of the society. Though they all want nothing but communality, individuals do not chose and join a society they create for their purposes, but must be ever "integrated" into the community by the community.

Set aside the logic of the odd idea, imagining the individuals of a society and the society of individuals as two separate entities, it is this odd image of a community splitting the individuals into competing privates only this community creates, into community members left alone without the community they are members of, it is this theoretical separation of the society of individuals into individuals without society, that generates *the* mission for sociological thinking to ever struggle for the re-integration of both, after only sociological thinking invented their separation, in the above case phrased as a historical process. Thinking the society members and the society as entities apart, failing to be part of what they are thanks to the society, members of a community, it is this paradox idea that constructs the need for a nation state's mission to ever integrate the individualized individuals as the categorical basis of sociological thinking.

The overarching sociological category for this false idea, the idea of "integration", of ruling, structuring, has several names, as do the sociological means to do this, may this be the "norm", values, institutions, or such things, it is always the same concept of all kind of communities structuring the otherwise disorganized life. Sociological thinking idealistically insists on a mission of nation state that is meant to regulate an otherwise dis-regulated life. They ever present the nation state's rationale and all the nation state entities and institutions as "communities" as to organize the dis-regulated citizens—the nation state mostly fails to fulfil.

Sociological thinking is to supervise individuals and communities to trace if they, the communities, are at risk. Communities are for sociological thinking the real concerns of humans and any politics, political or social institution, including nation states, must be observed if they question communities. Sociological thinking is thinking about the stability of systems, structures, values, in short anything providing the *'Ordnung'* of a community, it is in this sense the self-appointed community police.

Observing humans whose life consists of all kinds of "changes", sociological thinking is always concerned with the question if the communities succeed in fulfilling their *Ordnungs* mission. Thanks to the fact that the nation state organizes the life of citizens as their competition about exclusive "biographies", sociological thinking detects everywhere the risk of communities failing their mission:

> "What we experience in various and specific millieux, I have noted is often caused by structural changes. Accordingly to understand the many changes of personal millieux, we are required to look beyond them. And the number and variety of such structural changes increase as the institutions within which we live become more embracing and more intricately connected with another. To be aware of the idea of social structure and to use it with sensibility is to be capable of tracing such linkages among a great variety of millieux. To be able to do that is to possess the sociological imagination." [89]

Consequently, sociological thinking seriously thinks also war among nation states as a case that proves a failure of the ordering power of nation states:

> "Consider war. The personal problem of war, when it occurs, may be how to survive it or how to die in it with honor; how to make money out of it; how to climb in the higher safety of the military apparatus; or how to contribute to the war's termination. In short, according to one's values to find a set of milieu and within it to survive the war or makes one's death in it meaningful. But the structural issues of war have to do with its causes; with what types of men in throws up into command; with its effects upon economic and political, family and religious institutions, with the unorganized irresponsibility of a world of nation states."[90]

Sociological thinking about war has a sound understanding for their spectators, may they make money out of it, join the war in the military apparatus or die in it with honor, thus all contributing to the war's termination, all honorable ways to make this set of milleux "meaningful" for them, despite the unorganized irresponsibility of nation states.

---

[89] C. Wright Mills, *Sociological....*, p 10/11
[90] Ibid, p 9

For sociological thinking the social and its ever detected changes of its structures, only sociological thinking detects to justify the need of sociological thinking about the danger for a disordered 'Ordnung', values, norms, in sociological thinking, institutions ever challenge the nation state's communities to care about "re-organizing" the social, nation states mostly fail and even take the case of a war, in which nation states organize their whole society towards this mission, to accuse nation states wars as a case of an "*un-organized irresponsibility*"—a critique that even presents the most organized and most violent actions of nation states with the most obvious objectives as a failure of the mission of nation states to organize the whole world. Sociological thinking thus only proves that sociological thinking about the world's social is obsessed with thinking through their idealized mission of nation states only they attach to nation states as to ever create nothing but "order"—a sociological mission, that does not want to think about what nation state are and want, but a mission that ever ends up in the disorder inside nation states and in the world, a disorder that forever guarantees the need for sociological imaginations, imaginations about a world challenged to ordering the world towards any order.

The unquestionable mission of nation states for sociological thinking is to provide a kind of order among the individuals, just as if the nation state was a kind of social traffic police and it is the concern of sociological thinking to ever critically question if the rules, the structure, the community, the society or the system allow the citizens to follow the rules and thus to be part of the community that sets the rules. And within this critique the nation state for ever violates its mission sociological thinking attaches to the nation state.

Sociological thinking cannot be confused by the fact that the nation state institutions provide their service with their power monopole against the citizens inside nation states forcing them to do what sociologists believe they desire. Sociological thinking has perfected a mode of critique that accuses the nation state of failing to carry out a mission only sociological thinking attributes to it. It is the critique nation states may not really appreciate, as their distance towards sociological thinking shows; it is however this critique that ever transforms any criticism of the impertinences of nation state, including wars, into a plea for its genuine sociological mission.

"But it is not true... that man's chief enemy and danger is his own unruly nature and the dark forces pent up within him'. On the contrary: 'Mans chief danger' today lies in the unruly forces of society itself, with its alienating methods of production, its enveloping techniques of political domination, its international anarchy—in a word, its pervasive transformations of the very nature of man and the conditions and aims of his life."[91]

As in all these cases of sociological thinking it ever *reflects about "man's chief danger"* negatively, the failure of the *"methods of production, its enveloping techniques of political domination"* in front of its mission. However, the fact that sociological thinking ever arrives at this critique ever accusing the *"the unruly forces of society"* of ever failing in its mission to organize humans lives, does not urge sociological thinking to raise the question, that, if this mission ever and ever fails, it might after all not be the mission of the *"unruly forces of society"* to rule—not to mention that this sighing view of the unruly society is a conclusion the otherwise ever critiqued political elites could fully share and appreciate. If the mission, ruling the unruly, ever fails and therefore might be only an imagination only sociological thinking has, is no conclusion sociological thinking could ever consider, since this would posit omitting *the* essential of sociological thinking, the idea of blessing to ever strive for the ruling mission of nation states.

Political theory—political power for the politically disempowered

The reality of the neat idea of the sociological society that strives for serving the desire of individuals for communality, is the sociologically idealized nation state, a necessity of the competing privates, constructs of the nation state, using *and* excluding each other interests, the nation state by force gathers beyond their particular interests as citizens under the nation states objectives and his rules, ruling the lives of privates with their exclusive interest under the objectives of its power monopole. While sociological thinking blesses the forced community of nation states as the never substantiated ideal of a society, political theories add a more accommodated view on the nation state in the concert of disciplinary theorizing, accommodated to the real objectives of nation state, sociological

---

[91] Ibid, p 13

thinking likes to ennoble as a subject that serves humans desire for sociality.

It does not irritate and does not call for no further explanations if social science thought in political theories states that it is the *power* of nation states with a monopoly on power in the capitalist society that serves its citizens to pursue what the privates want:

> "At its simplest level the state is nothing more than formally organized government as we know it in the modern world. A state is a legally formalized entity having accepted jurisdication over a territory and a population and the capacity, within that territory, to make rules binding on the whole population and to enforce those rules through generally accepted legal procedures and applications of force. The state is an entity in which sovereignty—the authoritative capacity to govern within a country—rests. In this capacity the state defends and speaks for its citizens..."[92]

Why the nation state humans need power to force them to achieve what they want to achieve does not cause any confusion in the mind of politological thinking, because for political thinking it is due to the nature of humans' interests that they can never achieve any agreements among themselves about how to get on with each other, except the one and only one, the "contract sociale", they can very well agree, which is that they need a power monopole above and against them, a power monopole that forces them towards agreements about coping with their conflicting interests.

It also does not irritate politological thinking that the interests, they are not able to combine them towards any joint economic efforts and which therefore need the power interventions of laws against an ever threatening implosion of the nation state society, that these interests are also defined by these very laws, fueling the implosives of this society. Indeed, there is nothing in the nation state society and the life of its humans that is not defined by law and, in the first place, it is the conflicting interests institutionalized by the state as lawful interests including how to lawfully perform and achieve them, that *creates and domesticates* these explosives. If people do what they are told to do by the powerful laws they are forced into their conflicts, conflicts the very nation state sets into force and rules via the lawful actions of the nation state humans. Political theory never raises any question concerning what such politics are about which impose a definition of societal interests

---

[92] John C. Donovan. et al, (1993) *People, Power, & Politics,* Boston, p 19

they force the humans of this society to pursue them as conflicting interests *and* which lead them into those very social conflicts that call for the same power of laws to keep a society working *despite* of the politics that cause and set into force via their lawful definition of these interests the conflictual ways to pursue them.

The cognitive means of political theories to ignore the contradictions of a power serving the objects of power is their image of humans through which they construct political thinking of political science about the sphere of nation state politics, transforming the logic of a "despite politics" into the logic of a "because of politics". The image to arrive at the logic of "because of politics" is to imagine the political citizen *as it is defined by the political power without the political power*, more precisely the definition of the conflicting political interests through the laws of the political power without the political power ruling these interests, thus arriving at an image of a society that would end up in the anarchy which these anarchic citizens thanks to the political power interventions manage to get under control.

> "Why indeed should individuals subordinate themselves to an institution that reserves the ultimate right to discipline them by force? ....Anarchism continues to have great appeal...All that is required is that one accepts two premises: that humans are naturally good and that no formal public system of social control is needed to keep them from aggressing against another. But the fact remains that the weak of any society need to be protected from the predatory and strong."[93]

Yes, why, indeed? As if any "individuals" have ever been asked, political theory sets the ground for the need for power into the needs of the first victims of power, whether this is the "aggression" of other "stronger" citizens or of the "institution that reserves the ultimate right to discipline them by force" and attaches the need for a political power a social mission, not noticing that this, needing protection against power, rather argues that "the weak of the society" are the least who have any reason to "subordinate themselves" to the powerful institution, because they are the first and only ones the power hits.[94] Those who are *made* weak by the political power, are only weak thanks to and with the power of its laws and they are

---

[93] Ibid, p 22
[94] This—to present the power of politics as an offer against power—is the essential pattern of political ideologies justifying any political activities.

supposed to be those who need to be protected from the strongest, the dis-empowered protected from the political power by the political power? Constructing the weaker and stronger as a result of social relations in which the political power does not occur, just is if the construction of both was not the result of the very political power, to then introduce the political power as a protection for the weaker humans, is *the pretentious theoretical fault and the* life lie of social science political theorizing. The Human Rights Declaration of the United Nations[95] has fixed this essential logic of presenting nation states, the power monopole as protecting those the nation state has disempowered, as the nature of humans and as the view how the Post World War II world of nation states has to be interpreted. Just as the life lie founding political theories, this embryo of all nation state ideologies uses the argumentative trick to ignore that it is the nation state, that creates with its protection of private property the economically powerful and the economically powerless, dependent on being employed by the powerful to survive with a salary and disempowered by the power monopole the thereby powerless citizens, with the result, that all citizens lack any political power so that the economically powerful can enjoy to dictate the rules for employing the economically and politically powerless humans. To present the promise of the political power to not use its power beyond the rules for using it against those it has disempowered, rules the very political power defines and sets in force, is the cynical logic of the political power, political theory ennobles with the glory of a scientifically well-reasoned thought.

Thanks to this theoretical foundation of the political power, imagining the political theorist's horror vision of the "aggressions" among the lawful humans without the power of laws, political theories once and forever are liberated from raising any questions about the contradictions of a power serving the interests of the people it rules[96]—in the language of political theories therefore—smartly "governs", and can devote its thinking to critically cele-

---

[95] See: http://www.un.org/en/documents/udhr/index.shtml
[96] It is this vision of the non-governed anarchist citizens that also prevents the political theorists from concluding that the answer on his question, *"Why indeed should individuals subordinate themselves to an institution that reserves the ultimate right to discipline them by force?" (ibid, p22)* might be, because they are forced to do this by this very political power, political theories prefer to interpret as to protect the weak from the strong.

brate the question if the means, instruments and spheres of the political power to arrive at the ideal of political theorizing, that is creating a sustainable harmony between the anarchic interests of the citizens and *their* political power.

Once the state is theoretically founded by political theories as doing nothing but solving the conflicts the citizens only have thanks to the state, any sphere of politics, may it be economics, fiscalism, private rights, employment, family, science, and—above all—democracy and elections, where this ideal of a ruling power is set into force by its ruled anarchic vassals are subsumed under the assumptive view of political theories critically observing if and how politics and the whole variety of political mechanisms manage *the* idea guiding political theories, to control the anarchic nature of humans—in the first place the very politicians, who, of course, are from the same mind breed as those of political theories.

Psychological thinking—the mythologization of the mind

Unlike political science, which develops with its idea of an anarchic human that is hard to be domesticated, a certain understanding for the troubles the political power is ever confronted with while executing its power against the powerless, just like sociological thinking psychological thinking is utterly critical thinking:

> "In trying to solve the terrifying problems that face us in the world today, we naturally turn to the things we do best. We play from strength, and our strength is science and technology. To contain a population explosion we look for better methods of birth control. Threatened by a nuclear holocaust, we build bigger deterrent forces and anti-ballistic missile systems. We try to stave of world famine with new foods and better ways of growing them....We can point to remarkable achievements in all these fields....But things grow steadily worse and it is disheartening to find that technology itself is increasingly at fault. ...The application of the physical and biological sciences alone will not solve our problems because the solutions lie in another field....In short, we need to make vast changes in human behavior and we cannot make them with the help of nothing more than physics or biology, no matter how hard we try."[97]

Setting aside the fact, that psychological thinking drafts its critique and the challenges that a "we" is facing, may war and poverty be phrased from the view of the "we" of an imperial state power or

---

[97] B.F. Skinner, (2012) *Beyond Freedom and Dignity*, Indianapolis, Hackett Publishing, p 4

from the view of mankind on *"terrifying problems"*, psychological thinking without any further concerns subsumes under this "we" the most diverse things to stress in the first place the critical view psychological thinking has on the world's social, though enumerating War and poverty next to building "antiballistic missile systems" can only be enumerated from the perspective of the concerns of an imperial power only from which view all these are "terrifying problems" accumulated as challenges for ruling the world. Needless to say, that also psychological thinkers very well know that the global poverty has something to with the global business and that building "antiballistic missiles" aims at disarming the missiles of the enemy, so that both poverty and building missiles might not be a failure, but the result of very purposeful and very successful *"human behavior"*. Psychological thinking presents those threatening scenarios as *failures* not to think about these failures, but as failures which should illustrate a lack in thinking about the world and to motivate the need for psychological thinking. Psychological thinking does not want to think neither about war and nor about poverty. Psychological thinking only mentions both and the failed means to solve "our problems", last but not least as the failure of other sciences, to ignore seeking for the reasons for both in both and suggest to no longer reflect on war and poverty, but to shift the debate to what psychological thinking believes helps to understand and to solve any human problem and advocate that this, unlike all the other ways of theorizing, solution is provided only by psychological thinking, here more precisely by thinking about "human behavior". Human behavior, what and how humans are, psychological thinking explains why "we", also the national "we" building our missiles and our food industry creating poverty, as the whole "we" of human's, doing whatever humans do, psychological thinking subsumes all this under the same "fault" in human behavior, all these humans make when they do all such false things and when they fail, may this be war, may this be producing missiles and may this be failing to combat poverty, *"no matter how hard we try"* because "we „are humans and because "we" do not understand what humans are. *It is only* psychological thinking as a social science discipline that is thinking about and that knows what humans are and that discusses *"with the help"* of psychological thinking *"how changes in human behavior"* can be made—despite that is human's nature.

To reflect on the *"terrifying problems that face us in the world today"* unlike other social science disciplines psychological thinking suggests to ignore thinking about what the nature of all the "terrifying problems" is and suggests to instead think about the nature of humans, "human's behavior". With the odd logic that firstly any actions of humans are actions of humans and therefore must be due to the nature of humans and, with the implied assumption, that secondly, the outcomes of "human behavior" such as wars and poverty are non-desirable things for anybody and thus a failure, psychological thinking creates its peculiar way of disciplinary thinking—and might already terminate psychological thinking, once psychological thinking knows that all the "terrifying problems" must be explained as originating from human's nature, since, as one could conclude, if they result from human nature, then they are a natural part of human life.

However, for social science psychological thinking interpreting the *"terrifying problems"* as a result of human behavior's failures is *the* point of departure for psychological theorizing thinking about how to intervene into human behavior, into the "mental life" of humans:

> "Psychology is the Science of Mental Life, both of its phenomena and of their conditions. The phenomena are such things as we call feelings, desires, cognitions, reasonings, decisions, and the like.[98]

What do we learn about mental life from the "Science of Mental Life"?

Psychological thinking is about mental life, more precisely about *"both of its phenomena and of their conditions"*. As its religious predecessor, psychological thinking is about an instance, conditioning the human mind, setting the reason for mental life, and, driven by the secularized religious idea of the modern human to no longer tolerate his life as his fate, but in the enlightened version of psychological theorizing, psychological thinking is to identify what steers the minds of humans behind his will crafting his will, which in the first place means to disregard the will humans have as a mere executing agent of the psychological instance behind his

---

[98] William James, *The Principles of Psychology* http://psychclassics.yorku.ca/James/Principles/index.htm

will, such as "*plans, purposes, intentions or the other prerequisites of autonomous man*":

> "The task of scientific analysis is to explain how the behavior of a person as a physical system is related to the conditions under which the human evolved and under which the individual lives... We do not need to try to discover what personalities, states of mind, feelings, traits of character, plans, purposes, intentions or the other prerequisites of autonomous man really are in order to get on with a scientific analysis of behavior."[99]

Psychological theories are about this essential of the "modern"—to use a category from sociology—human, the mind of a human that does no longer consider his life as his fate arranged by forces beyond him but by himself. The "modern" human is bound to craft his life by instrumentalizing the other humans live aims at their cost for the sake of its own benefits.

Needless to say that such relations benefitting from the others loss is most conflictual and contains some abilities to cope with the brutalities of this nation state way of living, not to mention that these brutalities also imply that not only a few humans already within the capitalist metropoles do not even have any means to benefit from others, and not to mention even more than the majority of people on the globe who are far away from being bothered with the mental challenges of the brutalities of subjects competing about to benefit from each other.

Such distinctions are beyond the scope of psychological thinking, which, as all other social sciences disciplines, reflect on *the* "mental life" of *the* humans without any further distinctions of any humans—of humans in the metropoles as the of the nature of any humans as the very same mental life as such.

Psychological thinking is theorizing about how using the free will to prove in all their various thought that the will of the "autonomous man" is not articulated in his will and actions, in his "purposes" but—in the behaviorist version—that these purposes only articulate the conditioned will of the modern human. Theoretically cleaning humans mind from the will humans have and articulate, in other words, to deny that the will humans have is their will, is the first operation of psychological thinking to then, in the next step, "discover" what the mind instance "really" is that makes hu-

---

[99] B.F. Skinner, (2012) *Beyond Freedom*......p 15ff

man will what it is, in the behaviorist case, *"a physical system (that, MK) is related to the conditions under which the human evolved."*

> "If we ask somebody "Why do you go the theatre" and he says "Because I felt like going" we are apt to take this reply as a kind of explanation."[100]

The way the psychologist constructs his example discloses the argumentative logic of psychological thinking. It is not a coincidence that the visitor of a theatre insists as an answer on the imagined question from a psychologist on his free will without giving any further reason for his decision than his will, saying "I felt like going". Giving any reason that is more than his sheer will to do so would have forced psychological thinking to discuss these reasons. However, since it is precisely this sheer free will and the articulation of any reasoning why he does what he does, that psychological thinking is determined to question, it constructs the non-reasoned decision, more precisely the reasonless reason, the content-less will, that allows psychological thinking to impose *the* question psychological is interested, which is to construct and to ask what the real reason behind the reasoning of our visitor to go to a theatre is. Psychological thinking wants to know, what is the reason for the reason is, reasons humans have, they are concerned with the freedom of the free will to set reasons and aim at the detection of what in their mind makes humans will what it is. They are concerned with a mind to decide that must have an instance that makes humans decide what they decide, they aim at knowing the reasons behind the reasons allowing psychological thinking to know what sets the reasons humans will sets.

To find this, psychological thinking always in the first step, theorizing about humans will, abstracts from the substance of the will the competing subjects have and articulate, to then, in the second step, once the real mind has been negated as being only a articulation of the unknown mind behind the mind, transform the self-assertion of competitive subjects in front of others of the same kind, the social relations of society members into an articulation of a battle of the self with itself, a battle that for psychological thinking is a battle within the mind between the mind instances, psychological thinking induces into the mind, a battle in the mind system of humans only psychological thinking is able to detect behind the mind

---

[100] B.F. Skinner, (2012) *Beyond Freedom* ...p 12

humans articulate. It is this way that psychological thinking constructs a mind instance that is *in* the mind of humans, unlike the idea of an external god, *and beyond their minds,* as the instance that *causes* what humans want and what humans do.

The model human through which psychological thinking as a social science discipline is approaching social thought is the assumption that humans mind must be steered by something in their mind that is beyond human's minds control. Psychological thinking starts from the mystery of an instance, a human driving force that crafts human's mind, an instance beyond human's will working in the will of humans, that makes humans will what it is, a will that makes humans want what they want—the enlightened version of a god steered human mind.

The predecessor of social science psychological thinking about a driving force behind *and* owned by human's mind is religious thinking and the will behind human will directing humans mind is god. And, no doubt there is something divine in psychological thought as psychologists confess:

> "Aristotle thought there was something divine in thought, and Zeno held that the intellect was God. We cannot take that line today..."[101]

Despite the heavy controversies among psychological thinking, they all share as the essential assumption to enter psychological thought that there must be a force behind the will of humans that is responsible for why they do what they do, which can however no longer be god. May it be thanks to Freudian inner driving forces or thanks to the more totalitarian idea of behaviorists considering the will of humans as a reaction to his environment or simply thanks to the biology of the human brain, psychological thinking is not only about any expressions of the human's mind, but about finding behind this mind this instance that makes the mind do what it does.

Psychological thinking is about constructing a mind agency behind the will causing the will, in the case of behaviorists—as the mind as a result of an effect.

It transforms the ability to set grounds for what humans do into the monster of a natural necessity of causes and reasons and therefore imposes into the humans brain an instance that *makes* hu-

---

[101] B.F. Skinner, (2012) *Beyond Freedom*...p 11

mans want what they want, thus finding what they seek "beyond freedom".[102]

> "To work for peace we must deal with the will to power or the paranoid delusions of leaders; we must remember that wars begin in the minds of me, that there is something suicidal—a death instinct perhaps—which leads to war, and that man is aggressive by nature. To solve the problems of the poor we must inspire self-respect, encourage initiative, and reduce frustration."[103]

The logic of psychological thinking is as simple as false: war and poverty may not be the objective of the moral codex of humans, humans who do want wars and who do wars and who do create and who do use poverty, for psychological thinking cannot execute the objectives these *particular* humans have and therefore must be the outcome of the instance in the *nature of all* human minds, "instincts", which ever explain any human activity, that violate the very moral principles competitive humans create to condemn them, any unwished will, here war and poverty—they condemn, if they damage their own objectives, and present thanks to the help of psychological thinking as the result of the non-domesticated "instinct" towards the condemned will.

War thus becomes the—ever tautologically—explained result of the disposition to war "a death instinct", thus justifying the critiqued war as the result of the non-domesticated nature of any humans, its "aggressive" nature, and the remedy against the disposition for war in human nature, the theoretical fault to domesticate what cannot be domesticated, since it is the nature, calling for the obscure insights of psychological thinking. If war must be explained due to the aggressive nature or thanks to the instinct for empathy, depends on a distinction between the same military action as a war or as a humanitarian intervention, a distinction the human mind is denied to make due to its mind that receives its or-

---

[102] Not surprisingly, psychological thinking consists of as many schools of thought as instances exist that play any role in the formation and articulation of human will, starting from sheer invention of driving forces behind the will to the seeking in the of biology of mind activities. Other psychological schools of thought create for their theoretical purpose the denial of the will to be a will and to find an instance that causes the will, a mind instances they call unconsciousness, a consciousness humans mind owns without knowing it and thus establish human nature with an attribute and a category that only exists in psychological thinking.

[103] B.F. Skinner, (2012) *Beyond Freedom*....p 10

ders from the obscure psychologically constructed instances, the psychological minds of any few exceptional autonomous humans must be able to choose between their aggressive and their emphatic nature setting the distinction between what must be considered as the effect of aggression and thus a war or an effect of empathy and thus a humanitarian action.

Psychological thinking thus takes the objectives, if they consider a military option as a war or as a humanitarian action, a distinction they deny human's mind to create, from those humans who are setting the moral or political standards for what must be seen as war and what as a humanitarian action. It is not that easy to conclude that it is the political elite who must be the only human creatures enjoying the freedom of a will to decide what the rest of the humans can only repeat as the articulation of their instincts.

Poverty—tautologically—becomes the result of the will of the poor lacking "initiative", the not very sophisticated psychological version for insulting the poor as to be lazy, and a result of the will to feel poor, psychological enlightened *"problems of the poor"* against which psychological thinking recommends to *"reduce frustration"*, the scientific version of the insult of poor people, that they did not feel poor if they did not expect so much. Whether poverty must be seen as an articulation of a lack of "self-respect" or a matter of unbalanced "frustration", depends on the ups and downs of the economy and the actual social security policy priorities, both of which psychological thinking has enough affinity with to allow to always very well know which instinct they apply to explain the "problems of the poor".

Psychological thinking takes its sound understanding for humans from the national Zeitgeist, presenting the societal demands for humans as the inner conflictual human nature with its instances steering the mind.

An example might illustrate the utterly moral, national Zeitgeist judgements steering the theorizing of psychological thinking:

156   Chapter B: Categorical essentials of disciplinary thinking

> "Our age is not suffering from anxiety but from the accidents, crimes, wars and other dangerous and painful things to which people are exposed. Young people drop out of school, refuse to get jobs and associate only with others of their own age not because they feel alienated but because of defective social environments in homes, schools, factories."[104]

As if the psychological theorist was a priest, Skinner repeats what the US media present as the concerns US citizens can read in their daily newspapers: *"Accidents, crimes, wars and other dangerous and painful things"* are the contemporary moral judgements brave nation state subalterns must consider as their concerns of an ever betrayed honest competitor, from which psychological theorizing takes its criteria to choose from its reservoir of "instincts" to ever tautological explain "wars and other dangerous and painful things" and their causes as their accordingly activated dispositions for them, may this be, what made somebody really decide to go to the theatre, somebody who might have thought just to have fun and feel like going. Thus, psychologists indeed share something similar to the divine knowledge of priests, since they translate the moral codex of competing citizens, no longer into the religious commandments from an obscure power above humans, but into conflicts within humans' psychologically obscured mind, which only psychological thinking masters to decipher, since only they created their religious like myths about human mind.

Psychological thinking is religiously obsessed and engaged in exactly this single false thought, varied through endless examples and cases: to deny the will to be the source of what humans want and do and to explain what humans want and do as the result of any instance beyond human will, whether this is their unconsciousness, the impact of any instance on humans mind, their disposition as in the case of behaviorist psychology, or, finally, the nature of his brain, in which psychological theories from there on dig to find the freedom of will as a physical mechanism to assist humans domesticating their will towards the needs of a competitive society. This, the self-domesticated human, is the ideal that constitutes and guides social science psychological thinking.

---

[104]   B.F. Skinner, *Beyond Freedom* .....p 14–15

Essentials of social sciences disciplinary thinking

1. The common cognitive lie founding the categories of disciplinary thinking

Presenting the domestication of humans as a service for allowing them to be social creatures despite their un-social nature is what founds all disciplines. They all impose what is the nature of nation state creatures into the nature of humans and the nation state as a response to this human nature, thus all founding the categorical essentials of their distinctive disciplines on this cognitive lie:

Economic thinking presents the market economy and all its elements as a service to provide goods for humans. Economic thinking translates the never satisfactory growing growth of abstract wealth into the never satisfactory nature of human's demands, of humans striving against each other for their private interests, and imagine the battles among these never to be satisfied "wolves", only the private property ownership creates, without the regulative power of nation states directing these creatures to cope with an ever scarcity—creatures threatening the ideals of market economy via an economic anarchy and, thus, arrive at the nation state as a service regulating the market economy despite of this never satisfactory human nature in order to provide the humans with goods.

Sociological thinking presents the nation state as a "community" that always helps solving the problems, the members of this community only have thanks to all these services—an actual example may be the presentation of the social security system serving the security of human's life, though it only reveals that human life under this security system is obviously never secured. Sociologist found sociological thinking on their horror vision thinking of an unsocial nature of humans and imagine these unsocial creatures without "roles", "structures", "systems" or "*Ordnung*" as the fortunate intervention the "society"—a sociological translation of the whole of the nation state creatures, provides its otherwise anarchic members, thankfully guiding humans through an otherwise incomprehensible jungle of options for actions, preventing the world and its inhabitants from the anarchy of the free will of humans.

Political thinking presents the nation state as protecting the live of citizens against all kind of foreign political subjects, all citizens only encounter as foreign subjects due to their status as nationals—

most prominent example is the presentation of conflicts with other nation states only the political materialism of nation states creates as a protection of citizens in front of the enemies, only nation states are able to create. Political thinking imagines a world of competing subjects without the domesticating power of "governance", thankfully preventing the world with a monopole on violence from violence the very nation state establishes and domesticates, a violence political theorizing advocates and denies, because the ruled are allowed to elect their political leaders.

Psychologist can only imagine humans mind as a mind that struggles with the conflicts only nation state subjects have, struggling with their conflicts between what is morally allowed and what the individual objectives are, a conflict only nation state subjects have, who perceive the orders of laws as a moral mission of man, a conflict psychological thinking translates into the nature of human mind with different mind instances, minds that must be controlled by the "self" via a mind instance beyond and inside their will. Without such conflicting mind instances psychological thinking cannot understand, why humans do what they do, minds only psychological thinking does not understand, because it shares the struggles only nation state humans have, struggles between the moral values and the desires of a competitive subject.

Anthropological thinking takes the variations of the categorical grounds for the distinctive disciplinary theories, derived from variation of essentially the same human nature, as the methodological mission, to detect the multiplicity of the individual human nature founding the individual disciplines as all derived from a human meta nature, a meta nature about which anthropological thinking thinks about, about humans as such, about "man", man, who demonstrates in what man in anthropological thinking is, is what anthropological thinking inserts and detects in "man's" nature, a "problematism" of human nature, that calls for all the "problem solving" theories disciplinary thinking thinks about, problems disciplinary thinking interprets as a response to human nature, a nature that discloses its origin in the nature of the bourgeois human, the citizen.

With this way of theorizing, interpreting the social as a response to an problematism in human nature, anthropologists call in one or another way the "incomplete human nature", not only the most opposing activities are explained by the social sciences as the same case of the same nature, not only the whole reality in this way of theorizing is never scrutinized as what it is, but is interpreted as variations of cases expressing the ever same problematism of human nature, but also the most violent activities only nation states are able to execute, such as war, become in social science thinking, that thinks the whole social world as articulations of the images they make up about problematic human nature, social science thinking manages to interpret also war as a response to cope with the problematism in human's nature.

Substitutionally for all the social sciences' and the service their human's images of a "problematic" creature provide for they ways of creating social thought, the most profound experts, knowing what humans are and want, psychologists, such as Skinner, know this about war and poverty:

> "In trying to solve the terrifying problems that face us in the world today, we naturally turn to the things we do best. ... To contain a population explosion we look for better methods of birth control. Threatened by a nuclear holocaust, we build bigger deterrent forces and anti-ballistic missile systems. We try to stave of world famine with new foods and better ways of growing them... We can point to remarkable achievements in these fields.... But things grow steadily worse, and it is disheartening to find that technology itself is increasing at fault.
>
> ...What we need is a technology of behavior. " [105]

Concluding from the hardly doubtable observation of the "real fact" that war and poverty are *made* by humans and not by the nature that it must be the though the *nature of humans* and, hence, that the world needs a behavioral technology that encompasses this nature to domesticate the nature of humans, a supra human power over the human nature, such conclusions of some supra human thinkers require not only some ignorance about the rationales creating war and poverty, but a religious like obsessive passion to interpret any morally discredited "evil" in the world, humans are morally not supposed to do, as a matter of the one and only evil, the human's behavior.

This, seriously considering war and all the world dramas as a matter of a non-domesticated human behavior, is not only the way psychological thinking, but the way disciplinary thinking as a whole theorizes about the social, explaining the social as a matter

---

[105] B.F. Skinner, (2012) *Beyond Freedom* .....p 4

of human nature that calls for its control—with the advice of social sciences' knowledge.

No social science discipline would be able to construct the theoretical foundations of their particular disciplines without this image of humans that needs across all disciplines nothing else and nothing more but the knowledge of disciplinary thinking about how to domesticate its evil nature. The hint, that their image about humans might have something to do with that "real fact" that all humans by force are made private property owners and the conclusion, that this implies a world of people forced into a battle against each other, for most of them, the have-nots, a battle about their sheer existence, for others, the have-its, a battle about increasing what they own by exploiting the have-nots, this is certainly a conclusion that is too much thinking about what is going on in the social world, a social world constructed as a world forcing humans into a battle among competing subjects, instead of reading the social world as an articulation of religious-like insights into the human nature.

There is nothing but the idea of free humans, the scientific propagandists of the free humans seemingly are so horrified about, only their false thinking detects as mankind's burden of human's nature. Social science thought developing perspectives for the lives of a burdened human that must be carefully observed by social science theorizing—discloses a very Christian view on mankind, disclosed in its secularized version populating the social of social science thinking.

2. The shared metaphysical nature of the disciplines and their speculative way of theorizing

Substantially, thinking in the different disciplines is not as diverse as the different disciplines appreciate to present their individual approach as a unique perspective to think about the social.

All disciplines basically share the same image of humans, the threat of an anarchic, non-domesticated, ungovernable human, a threat only disciplinary thinking creates thanks to the cognitive affirmative thinking of the social science approach to social thought. Imagining the competitive humans, the very creatures of nation state societies, imagining these very particular humans only this society system creates, acting as a threat for each other, if not controlled by the "ordering mechanisms" of anthropology, controlled by the "society" in sociological thinking, by the political power in policy science and via its self-domestication in psychological thinking, all these disciplinary constructs are invented by the individual disciplines and their slight differences found the way of theorizing in the individual disciplines about the social.

All the variations of an image of an anarchic, non-domesticated human, created to prove the need to get their own anarchism controlled by any instance, these anarchic creatures in a mystic act of anti-anarchism have created opposed to their nature, in sociological thinking this mystic image of humans grounds the need for a "society" which as opposed to the members it though consists of, this "society" is an entity that gathers all these anarchic individuals in a "community", a "community" that has the mission to get the nature of its members domesticated. It is a construction of a "community" that not only denies the politically established and supervised conflicting interests among its community members as created by the "community", it is this idea of a community compensating the nature of its members that also negates the political violence needed to set the abstraction of equal individuals into force, imagining these creatures, the very "modern" society *makes*, in which they are forced to using each other's conflicting interests as a means for their privatized life aims. It is the achievement of the social sciences across all disciplines to transform the creature only this society model creates into the scenario of a human nature, imagining individuals as wolves without a domesticating leader. It is this fairy-tale like vision of a non-domesticated nature of humans that constitutes the variations of the disciplinary thinking across the disciplines, obsessed about a vision of a very human only this society model gave birth to and only disciplinary thinking creates as the nature of humans, thanks to their affirmative way of theorizing, that does not want to see the "facts" they so much advocate to think about, but derives its theoretical foundations from images they though only receive from the society they reflect on.

Strictly speaking social sciences do not really think about the social world, but the real facts freaks instead think about if and how the world mirrors the imaginations social sciences have about the social world, just like religious thinking derives from its images about the world that *the world is how the world is supposed to be.*

Chapter B: Categorical essentials of disciplinary thinking

In disciplinary thinking, not the social world is analyzed how it is, in disciplinary thinking this is all just the other way round: It is the "community", the "market", "governance" and the "self-control" of humans will, which all domesticate the human—social scientists would say, a set of social constructs, social facilities, which orchestrate their shared mission, that is to allow humans to strive for their life aims despite of the obstacles of human nature, ever critically observed by social sciences, observing the imperfect ways in which all those social constructs, starting from the "society", over the "market" towards the "self", critically observed by social science thinking if and how they manage to make the social world to coincide with their images about the world and this image is that the social is a means serving humans to be social humans.

The fact that these modern societies, created and ruled by nation states, allow their citizens to strive for their life aims as privates, misleads social sciences to think about the relation between the political power and the citizens other than of the nation state as a service for citizens. It is this permission to strive for private life aims that creates the illusion founding social science theorizing that it is the objective of this society to assist the privates to do so. Thinking that it might be just the other way round, that the privates are the means of the nation states in their battles about power among nation states, objectives nation state impose into humans life via laws as ways and means to pursue any private life aims, ever pretending that they were only framing the life aims of humans via the legally established rules, while these so called life frames impose the nation state's rationales into the life agendas of their citizens, understanding the relation between nation states and citizens that it is the nation state which makes its citizens his means for his objectives, is unthinkable for theorizing that cannot think about the nation state social other than a service for its citizens. Wherever this mission does not coincide with the reality, then, for social sciences, it must at least be its genuine mission. Thus, theorizing about the social, thinking about the nation state social beyond the nation state service paradigm, is unthinkable for the social sciences, since their categories across disciplines are constructed on this illusion, interpreting the permission to strive for their private life aims as if permitting to strive for them was the same as serving them to pursue them.

3. Disciplinary social thought cannot think other about the social but as an idealized nation state social

As a consequence of thinking nation states as compensating the un-social nature of humans and as the only means allowing humans despite their nature to be social creatures, social science theorizing, thinking through the categories founded on this concept of humans and of their nation state, is conceptually unable to theorize about the world's social other than through the perspective of these constructs of idealized nation state missions, disciplinary thinking attaches to the nation state and its creatures as a necessity of their nature

Translated into their constructs, such as the "control mechanism" in anthropology, the "structures" in sociology, the "governance" in political thinking, the "self-controlled mind „in psychological thinking, all constituting their according disciplinary categorical apparatus, through which they theorize, all these theoretical constructs are the idealized images of the real nation state constructs, which not only constitute the theoretical essentials of each discipline, but which for social science thinking make humans social.

Since humans are only social due to these nation state constructs, disciplinary thinking attaches to the human as only thus creating their social nature, social sciences cannot think other about the social, than thinking the social as these idealized nation state constructs. Social sciences not only reflect about the social through the perspective of these nation state constructs, all the various forms of citizenship, such as a student, a worker, a private property owner, a retired person, children and a tax payer, social sciences not only reflect on the social via the constructs disciplinary thinking invents from their images about an non-yet-social-human, images that found each discipline's category system, these invented nation state constructs of disciplinary thinking constitute for disciplinary thinking the sociality of humans, so that it is the—idealized—nation state, which via these constructs only generates and guarantees the sociality of the otherwise non-social human nature. In short: For disciplinary thinking, their idealized nation state missions make the human a social human, for disciplinary thinking it is the nation state and his constructs which generate the social and, hence, generate what *social* science thinking is all about, thus give birth to the social sciences, which were therefore during their emergence in this fundamental sense rightly called "*Staatswissenschaft.*" Without their idealized nation state creatures there is no social, is the assumption on which social sciences thinks. It is the disciplinary interpretation of what the nation state is supposed to be, what the social is, about which the social sciences think.

No doubt, it is the merit of social science thinking, to put social thought from "head to feet", though only in a very formal, rather mechanical sense of its cognitive operations, however, they anchor their reflexive feet so fundamentally in ideals they impose into the human as their human nature, they though only receive from the reality, ideals which show that they are constructed from a variation of very religious like thinking, the bad human ever rescued by a "deus ex machina", a god like nation state—however, only god—*like* and therefore ever a matter of the very critical observations of social science theorizing.

4. The categorical essentials: Critically affirmative and idealistically domesticative

Indeed, social science thinking is by no means a mere apology of the nation state or any real social. For social sciences the real social is only an ever imperfect realization of what they believe is its disciplinary wise distinguished mission, fulfilling the challenges to implant and foster these missions into human life. Moreover: The social reality is not even what disciplinary thinking thinks about, the real world is only a "fact", facts which must prove that the world really is about the metaphysic missions disciplinary thinking derives from their metaphysic images they construct about humans and, metaphysic missions, against which they measure the real world as aiming at and serving these missions. The world on which social science reflect on is only a world of their metaphysical ideas, the real world is only an ever non-perfect realization of their ideas, they ever critique for mostly failing to match with their ideas about the social.

Anthropological thinking interprets the social world as articulations of their image of an incomplete human; sociological thinking everywhere seeks for risks of changes, of structures, or systems, of relations, all creating and securing the otherwise lost sociality, all social mechanisms, ever risking their mission to create sociality;

economists calculate and calculate figures, cross-checking any violations of scarcity, the negative phrase for growing growth, for their "never enough"; political science is concerned about the governance of the anarchic human and critically observes policies if they manage to cope with this mission; psychological thinking interprets any articulation of any will as indicating the troubles the self has with its mind instances and feels challenged to critically observe if the human troubled with the conflicts in his self is over—or not enough burdened, handling his troubled mind.

Thus, disciplinary thinking combines critically affirmative and idealistically domesticating knowledge by presenting the domesticative means of the society system as helping a human to get on with his nature, critically observing if the means of the society provide this service to its citizens.

Indeed, social science thinking reflecting on the social compared against this idea of a social serving their citizens compensating their un-social nature, this very idealizations of the social in social science thinking deviates from the real rationale both of the economy as of the nation state and social science theories are therefore never fully appreciated by the representatives of both the economic and the political elite. It is this difference between the idealistic rationale of disciplinary thinking and the real rational of the materialism of politics and of the economy that causes the distance and tensions between science and politics, which in some cases even results in massive conflicts.

Both the political and economic elites appreciate social thought, as long as they interpret the discrepancies between the promised services of the state and the economy as failing to provide the promised service, which therefore call for further political and economic actions and thus help to reinforce these ideals of a promised service. Political and economic elites also appreciate to be critiqued against these expectation social science sets into force, that is that it is the mission of the state and the economy serving the citizen, and if they do this the more this mission fails; however, they only appreciate this as long as they insist on these very ideal missions.

If disciplinary thinking, however, insists so much on the mission they attributed to politics and to the economy, that they start to question if the particular political and economic elites really share the ideals they attributed to politics and to the economy, some clarifications by the latter clarify where the differences between a very welcome scientific idealism of social science thinking and the rationales of politics and economics are and from where on social science knowledge is denounced into the world of wishful thinking,—, "only theories" one may share as a scientific thinker, but social thought that is beyond any rationales that must guide the practices of a political or economic practitioner.

It is this difference that makes social sciences ever complain about never being heard, despite of being the societal representative of scientific expertise by the political and economic elites, who do indeed have another rationale than social science thinking wants to believe.[106]

Hence, unlike in the capitalist metropoles, it is in some countries of the "third" world not unusual that social science thinking are confronted with the massive opposition of the political and economic elites, if they consider that the confrontation of the idealized rationales of disciplinary thinking and their critical engagement for their rational of a concept of "society" serving their constructs of humans is too much revealing that these ideals are not the ideals of any political practices and are thus undermining their own political ideologies, ideologies, for which they use theories from social science thinking, but use them to prove what is not the same as what social sciences prove. Generally, this difference between the ideologies for the materialism of political and economic rationales and the critical idealism of disciplinary thought are, however, for the political and economic elites a '*quantities negliable*', as the institutionalization of social sciences as a state affair and the employment of all the critical thinking academia in a both state and private run universities demonstrate.

5. The world's social in disciplinary thinking—absent

Thinking about the world's social through the disciplinary models of humans, nation state constructs and their made nature, implanted by disciplinary thinking into the nature of humans, models through which disciplinary thinking interprets the world's social,

---

[106] Typically, social scientists who are working as policy consultants, appreciate to present social sciences as "subversive": "...Anthony Giddens argues, that sociology is an inherently controversial discipline, one that has a subversive character." Giddens, A. and Philipe W. Sutton, (2010) *Sociology: Introductory Readings* (3rd edition), Cambridge, Polity Press, p 2

do not allow social science thinking to think about the world's social other than a mere agglomeration of the many nation state socials, a world's social, consisting of a multiplicity of national biotopic socials, matching with or deviating from their modelled ideals about the social. More precisely phrased, thinking about the world's social for social science thinking means to reflect on the world's social as the many variations of nation state socials as the many interpretations of how these interpretations of nation state socials comply with their idealized missions, missions, which make—along the different disciplinary models of humans—these humans real social humans from those humans, burdened with an un-social nature. Since social science thinking is to measure the real social against their ideals of a social, ideas, according to what the social is supposed to be, materialized in their disciplinary categorical essentials, thinking about the world's social in social science thinking *is the same as* looking at the nation state socials as *ever comparing* their ways of *complying with or deviating* from the missions disciplinary thinking attributes to them as their genuine rationales, all executing the missions disciplinary thinking detects in the challenges of the burdens of humans nature, into which they have transformed the challenges, only the nation state socials know.

Interactions among nation states, the real forces crafting the world's social more than anything else, do not occur to the view social science thinking has on the social, a view that reflects upon nation state only as a practiced interpretation of what social sciences consider as the nation states objectives, such as to provide ordering mechanism for the incomplete human, to create and serve structures for the disorientated humans, to control the laws of scarcity, to govern the ungovernable and to allow humans to find a balance within their conflicting mind instances.

The world's social and its impact on the individual national social does not only not exist as a topic in social sciences. This, not seeing the interactions between nation states as the major forces, crafting the world's social, does not only apply this view to all nation state socials, to those which are ruled by others and vice versa. The topic of disciplinary thinking is to reflect on the social a secluded national social, just as if the world beyond did not exist, or, more precisely would only exist as other secluded national socials beyond of the same kind of isolated social entities. Their theories theorize about these secluded national socials by comparing these national socials against these genuine missions of nation states. "International policy affairs" are therefore the topic for sub-departments of economy and political sciences, departments of social sciences, which are not considered as genuine social science theorizing, but rather as policy consultancy.

To conclude from the fact that the life of humans is entirely dependent on the nation state humans are citizens of, that it is only this nation state that crafts their lives and to therefore exclude the world's social as a topic of reflections, is the essential false conclusion founding social sciences thinking, a false conclusion that is responsible for how they do—essentially not—theorize about the world's social. The world's social and the world's nature as a whole—the globe and its inhabitants, for nation states the resources of power they aim to rule, nation states striving for power over other nation states to use the world's people and nature beyond their ever to narrow territories they rule, the economy and the political rationales crafting human's live on this globe, is simply not a topic for social sciences.

And what is the purpose of all this, ruling the world beyond the ever to narrow national territories? It is the purpose of nation states to gain more power over other nation states, a purpose which is as abstract and violent as the objectives of the economy, aiming as growing growth, an economy nation states establish and supervise as their very means towards their purpose, gaining the economic means of the world for more power. It might be the abstractness of such purposes which prompts social sciences to invent others.

Chapter C
The social science approach to scientific thinking—advancements of teleological theorizing

Across all disciplines social science thinking practices a peculiar approach to theorizing, particular cognitive technics creating thought. The social sciences live in a world of ideas they construct from their metaphysic imaginations they have about the human nature, imaginations which are though only translations of the nature of the very nation state creatures they impose into humans as their nature.

"Scarcity" in economic thinking translates the growing growth, ruling a capitalist economy, as the missions to ever strive for an abstract, measureless *more* as the nature of any economy. "Structures", "relations" and alike in sociological thinking translate the forced subordination of the individual under the rules of the nation state and its facilities as helping the individual to find what they—due to their lacking sociality—are all striving for, relatedness, sociality. "Governance" in political theories translates the subordination under the political power as compensating the anarchic nature of humans. The struggle between the mind instances in psychological thinking translates the struggles of the individual between the precepts of morality and the interests of the individual as a natural conflict within a split mind. "Ordering mechanisms", "rituals" and, later, "culture" and the like in anthropological thinking translate the subordination of humans under social regulations and rules as complementing the lacking abilities of humans to become though humans despite his incomplete nature.

Their mode of theorizing is to think about the social reality as *comparing* the social reality against these missions disciplinary thinking attributes to it.[107] Social science theorizing practices a particular mode of comparative thinking, that originates from a mode

---

[107] Certainly, the nature of any thinking is to compare the known with the unknown. Disciplinary thinking, however, compares their object of thinking against their presuppositions through which they theorize.

of cognition that is epistemologically comparative in the sense that it approaches its object of thinking through assumptions, assumptions which consist of the missions they attribute to their objects of thinking to find out, if the "facts", the reality, coincides or deviates from what they are supposed to be.[108] Social thought under the regime of social sciences therefore ever is a cognitive journey between an affirmative realism and a critical idealism.

It is not that demanding to notice from the categories founding thinking in the social science disciplines that the way this thinking constructs its thought is guided by the determined will of the thinker to interpret what appears as the living conditions of people as a useful means for the very individuals, nation state societies set free and thus seemingly make the life agenda of humans a matter of these humans. It is neither nor: the fact that humans are the ones who decide about their lives within the given social constructs of nation state societies, allows to create this image as if the nation state constructs are a means of humans, thus for social sciences confirming the mission they translate into a mission of all the nation state constructs generating and serving the social humans, humans who are not able to create due to their un-social nature. It is rather the other way round than the ideals think, ideals found in the un-social nature of humans, ideals that not only found how the social sciences theorize across all disciplines about the social, but also establish the way they theorize, the technics of social science cognitions as comparing the social reality against these ideals.

As the above discussions about the categorical foundation of the social science disciplines show, the disciplinary way of thinking reflects on the social through a given set of assumptive modelled thoughts and interprets the social through a set of disciplinary constructs of the social, which represent ideals of the ways human are practically constructed as the distinctive subjects in the metropoles of the nation state society system, distinguished in ideals of the private individual (Psychology), the citizen (sociology), the private property owner (economic theory) and the political citizen (politi-

---

[108] "Empirical research" has developed this way of theorizing with its methodological apparatus towards a kind of technology of thinking. It is however, not at all only empirical thinking that practices this mode of cognition. This mode of cognition is the necessary way of theorizing, which thinks about its topics as comparing them against assumptions through which they approach them.

cal science), subjects disciplinary thinking derives from the religious like images they find in the nature of humans. These ideals of the distinctive humans in the metropoles of a nation state governed society system, constituting disciplinary thinking and the conceptual apparatus through which disciplinary thinking approaches the social, share across all their disciplinary categories to present the social reality as useful means for the society subjects responding to the problematized nature of mankind

Thinking as comparing the social with its ideals, is *the* social science way of thinking that establishes a kind of speculative knowledge, developed from their disciplinary images of humans, developed with a monstrous apparatus of methods proving or disproving the assumptions of their speculative thought against their coincidence with the objectives of a reality, ever tautologically constructing this proof as a comparison, compared against a reality constructed from their images about the reality, they therefore call the "empirical reality", a notion that perceives the reality as the real reality as opposed to their ideals against which the reality is reflected on. Taking the reality as a cognitive vehicle proving or disproving their speculative assumptions, social science thinking is thus methodologically affirmative, a mode of thinking that introduces the real reality as a means for judging about what knowledge is and what not.

One could but does not need to read all the more or less sophisticated books about knowledge and science to understand that there are different approaches to theorizing and that it is the particular format of thinking that directs the thinkers mind through a process of thinking about the object of thinking that results in different thought. Religiously inspired thinkers, let's say thinkers from an Islamic social science approach, certainly not only create different thought about the same object of thinking, different than non-teleological thinkers, and they do this, due to different cognitive relations between the thinker and his object of thinking.

Far before such obviously different ways in which thinkers from different disciplines approach their object of thoughts with most obviously different categories, scientific thinking in the social sciences across all the disciplinary categories practice a particular cognitive way approaching their object of thinking to arrive at the knowledge of disciplinary thinking, the disciplinary knowledge discussed before. Social science do this by approaching their object practicing thinking finding out if their therefore constructed reality is coinciding or deviation from the missions or objectives they implanted into the object of thinking as their responses to images social sciences detect in the human nature.

As a consequence of this way of epistemologically comparative thinking, it is the particular technic of cognition through which the social sciences thinker's mind penetrates the un-known object of thinking that underlies scientific thinking in the social sciences and that—necessarily—creates knowledge that combines idealistic and affirmative thinking, a methodological critical affirmatism, sketched in the above observations about the categorical foundations of disciplinary thinking founding the multiplicity of social *sciences*.

Unlike the considerations that follow, analyzing how the social sciences construct their thought, analyzing what the particular cognitive relations between the thinking subject and his object of thinking are, the social science technics of teleological theorizing, thus finding out what the particular nature of how social sciences think is, is most different from what social science epistemological theories discuss as varies " approaches" to social sciences. Under the notion of "approaches" of social science theorizing, such as positivism, critical rationalism or interpretivism, to mention only a few examples, under this notion the social sciences carry out epistemological reflections about different ways to operationalize the self-fabricated dilemma of a way of theorizing that—as a consequence of a false critique of the idealism of the classical philosophies—insists that knowledge is the creature of the actions of the human mind, of thinking, but that thinking that must not only simply analyze reality, but thinking that must attribute to the object of thinking the cognitive role to be a proving instance for the thought thinking creates.

What social sciences discuss as the different "approaches" to social science thinking are variations of the substantially same attempt, all trying to handle the self—fabricated contradiction that thought, the products of cognitive operations of the mind, must be found as being voiced by the object of thinking, that though does only disclose, what it is, through thinking. With their false concept of objectivity, interpreting the objectivity of thought as voiced by the object of thinking, social science theorizing struggles with the self-fabricated problem that theorizing proves ex-ante assumptions, materialized in theories as—in this sense of objectivity—objective theories, objectivity as the prove re-finding their presuppositions in the reality via their approaches to theorizing that approach their object of thinking with any ex-ante theories, the object of thinking must prove to imply and therefore, to handle this cognitive contradiction, construct a variety of cognitive operations, called approaches to theorizing, a set of cognitive mechanics, technics of thinking, procedures to operationalize this contradiction. Historically the last version operationalizing this contradiction is the dissolution of this contradiction via the dissolution of what constitutes scientific knowledge, so far at least aiming at such a kind of contradictory objectivity, the dissolution of even this contradictory concept of objectivity by the post-structuralist ideas, an "approach" advocating the epistemological absurdity of an "insurrection of subjugated knowledges"[109], advocating a multiplicity of subjectively objective knowledge, the very "approach" that provides the epistemological grounds for abolishing science for the sake of authentic spatiological theories. (See section A in this book.)

The social science mode of thinking—cognitive operations of a methodological idealism

The particular mode of scientific theorizing in social science thinking can be studied along any social science thought. It is done here along a rather arbitrarily chosen and very instructive example from economic theorizing, reflecting on an essential insight of economic theory, the equitation of demand and supply in the market economy.

This economic theory is presented under the following book title:

"Macroeconomics and Reality"[110]

In the first step of the following reflections tracing the nature of social sciences thinking I will analyze this notion in the book title on-

---

[109] M. Foucault, (1976), *Il faut defendre la societe, Cours au College de France,* Paris, Edition Seuil, Gallimard, p 8

[110] C.A. Sims, (1980) *Macroeconomics and Reality, Econometrica,* Vol. 48, No 1 (Jan. 1980), p 1

ly with regard to the way this economic thinker tells us *how* he constructs his thoughts, that is how his mind penetrates his object of thinking to discover its unknown nature. Then, in the second step I will discuss how this mode of developing insights affects the content of the knowledge he constructs through this way of theorizing, his insights about the economic relations of demand and supply.

Regarding Sim's mode of thinking, his technics of cognition, the first observation one can make is that what he thinks about, the economy, in his way of theorizing, seemingly exists twice: Assuming that macroeconomics is a theory consisting of insights about the economic reality, in other words, that the macroeconomic reality is the substance of thought he discloses in his macroeconomic theory, in his theory about the economic reality, this reality, transformed in his theory from an unknown phenomenon into a set of thoughts, appears in his way of theorizing a second time.

In this duplicated appearance his object of thinking, the macro-economy, he already theoretically possesses in his mind as macroeconomic theory, that is, as what he knows, what macro-economy is, the object of thinking that has lost its alien otherness by making it a thing the thinker knows, re-appears as an insightful instance, more precisely, the macro-economy re-appears in its abstract status, the reality of what he already explained, now as an instance, as an agent of thinking[111], a benchmark to judge about what his theory about macroeconomics is, a theory or not. Sim's macro-economic theory, one must firstly conclude, is obviously a theory, he seemingly already had about the macro-economy before he approached his object of thinking. More precisely, his insights about the macro-economy are insights, which he already had before thinking about the macro-economy. Put in other words: The particular mode of thinking about the objects of thoughts is to transform the unknown object of thoughts into knowing with thought the thinker already has before he thinks about what the object of thinking is and it is the object of thinking that is both the object and an instance, a cognitive driving force of thinking, a means for thinking about

---

[111] In fact, some departments of psychology consider thinking as a reaction of the object of thought on the brain and are searching in the human brain with the means of natural sciences this philosophically constructed metaphysical thinking agent.

thought, thought the thinkers had before thinking about the object as another means of thinking.

This raises three questions: Firstly, what the object of thinking as a cognitive instance in the process of constructing knowledge is, a cognitive instance the thinker attributed to the reality he reflects about next to being the object of thinking. Secondly, if theorizing means transforming the unknown reality into a known reality, approaching the unknown object of thinking with thoughts the thinker already has before he knows what the object of thoughts is, raises the question of the origin of these thoughts and how these thought he had before thinking about the macro-economy intervene into the process of thought creations.

Answers on both questions can be analyzed from a piece of thought constructed under the above headline, that is looking at the macro-economics, thought that exists before thinking about the macro-economic reality to develop thoughts about the macro-economy through these already existing thoughts.

> "In principle, we realize that it does not make sense to regard "demand for meat" and "demand for shoes" as the products of distinct categories of behavior, any more than it would make sense to regard "price of meat" and "price of shoes" equitations as products of distinct categories of behavior if we normalized so as to reverse the place of prices and quantities in the system. Nonetheless we do sometimes estimate a small part of a complete demand system together with part of a complete supply system—supply and demand for meat, say. In doing this, it is common and reasonable practice to make shrewd aggregations and exclusion restrictions so that our small partial-equilibrium-system omits most of the many prices we know enter the demand relation in principle and possibly includes a shrewdly selected set of exogenous variables we expect to be especially important in explaining variation in meat demand (e.g., an Easter dummy in regions where many people buy ham for Easter dinner)."[112]

To better understand the mode and the process of thinking, the ways in which this Nobel prize thinker is developing his thoughts within this above duplication of the object of thinking into the substance of thoughts and a means of thinking, an imagined agent creating thoughts, it is helpful to understand the contents of what is said here about the equitation of supply and demand, a topic in economic theory that, as anybody who is only slightly familiar with economic theories knows, is essential to economics.

---

[112] Ibid... p 2

What he is saying, is, briefly summarized, this: Firstly, he knows that demand for shoes and meat are not the same. They are, as he phrases it, "*products of distinct categories of behavior*". People, who buy meat, buy meat and not shoes for peculiar purposes and neither shoes nor meat can be replaced by each other nor can their use for which they are bought. That is, that the demand for shoes and meat are disparate things, not the same. It sounds odd to say this, but, as we will see, it is exactly this, that shoes and meat are the same, the economic thinker will argue that this is the case. Secondly, the prices paid for shoes and meat might also not be the same; however, with the little help of the idea of making "*shrewd aggregations*", one can say "*that it does not make sense to regard "demand for meat" and "demand for shoes" as the products of distinct categories of behavior*".

This logic is remarkable: If we assume, thanks to the idea of "*shrewd aggregations*" of the *prices* paid for shoes and meat, one must conclude that the *demand* for the things we buy, shoes and meat, are the same. In other words, if one assumes that pricewise demand and supply are the same, one can conclude that demand and supply for these particular goods and their particular use are the same. Since one can add together the—shrewd—quantities of money to be paid for buying shoos and meat, on can conclude that if people buy shoes and meat they not only buy the same but that both meat and shoes, from this economist's view, *are* the same. Quod erat demonstrandum: The economist equalizes demands for unequal goods and thus, by aggregating their prices, proves the equality of demand for shoes and meat. How about the idea that many more or much fewer people have a demand for shoes or for meat, but do not wish to pay them, for which reason ever?

To return to the way of constructing thoughts and to the role exante thoughts play in constructing thoughts and the cognitive service the duplication of reality in objects of thoughts and a means of thinking provides for thinking.

Sim's thinking operations analyzed above along with his theory on the equity of the demand and supply are to cognitively re-model the phenomena of reality and interpret reality by adjusting the reality to the way the reality exists in his ex-ante thoughts, a model, and presupposition about the object of thinking he had created before thinking and through which he approached his object of thinking. Shoes are no meat and demand for shoes are no demand for meat and cannot be made the same. He though does exactly this, equalizing both with his little trick of "shrewd aggregations": Arguing that the exact amount of the prices of shoes and meat could be considered as the same since the sum of their prices equal shoes and meat—he though admits are also not the same—he argues they can though be considered as being the same thanks to his very free cognitive operation of shrewd aggregations, so that one could rightly arrive at the equity of the demand for shoes and meat—exactly what Sim's set out to prove. This cognitive operation, transforming the two different qualities of the two goods into the same quality as both representing amounts of money which can be aggregated towards the same sum of money—if one ignores their different amounts—simply presupposes the very sameness of the two *products of distinct categories of behavior"*, which he wanted to prove through his cognitive operation. In other words: Sims cognitively re-modelled the two different goods into the same quality by—falsely—identifying their diverse qualities, once as qualitatively different goods rather than as prices to prove that their quality as goods are the same because their prices are the same and can be therefore added to quantitatively equal units. Being a macro-economist, not being interested in shoes or prices but in the equality of prices, he reconstructs his duplicated reality with those presuppositions, his ex-ante theory presupposes, here the equality of demand and supply.

One can conclude: The role the second version of the duplicated object of thought plays in his cognitive operations is to re-model reality in such a way as to allow to prove—or disprove—the modelled theory and its presuppositions through which the thinker approaches his object of thinking. In social science thinking theorizing means thinking in order to prove a theory social science thinkers already have before thinking. Re-shaping the objects of thinking towards proving presupposed thoughts is the social science way of thinking developed to a sort of cognitive professionalism in the social sciences, a cognitive professionalism that gave birth to a new epistemological category: The *empirical reality,* representing the duplicated reality in social science theorizing, is the modelled reality, modelled through and for the presuppositions of the ex-ante thoughts through which they prove or disprove the assumptive thoughts via what Comte named with another revealing duplication, the "real facts", the cognitive element of a cognitively constructed reality, constructed for this particular type of social science thinking one may therefore describe as teleological thinking.

As the example of "shrewd aggregations" also demonstrates, it is not at all only social sciences practicing what they call "empirical research" which practice this mode of thinking that proves thought created through pre-modelled thought via a comparison with the empirical reality, a reality constructed through these very modelled thought.

Professionalized thinking has further developed this epistemological categorical embryo of teleological social science thinking towards a whole language only the inaugurated druids of the social sciences masters are able to understand. The monstrous apparatus of methodologies is constructed for this single monstrous circular thinking about the real reality, modelling a reality towards an empirical reality for and through which social science thinking reflects on its presupposed ex ante thoughts, a world that only exists in social science thinking, again, not at all only for what is called "empirical" theorizing.[113]

As Sim's way of theorizing and his example of "shrewd aggregations" also demonstrates, it is this idea of an empirical reality, the scientifically constructed reality, constructed through the presupposing theories through which social sciences reflect and create their theories and against which they prove or disprove via comparing their theories against this constructed reality, that must have inspired thinkers like Foucault and other contemporary philosophers to the false conclusion to then finally radically omit the idea of any objectivity, even of the existence of any object of thinking in the sense that the whole world is always only what thought makes from the world, an anti-scientific scientific hubris that denies the objectivity of a world beyond the thought thinking constructs from it, an anti-scientific scientific hubris which represents the latest social science theory about social science theorizing.

---

[113] Just as if the repetition of a false thought was the elimination of a false thought the methodological apparatus has ennobled circular thinking towards a "hermeneutic circle" and perfected the repetition of this thinking in circles towards methods such as the method of "Grounded Theory".

Social sciences theorizing about social science thinking

Social science epistemologies know about the cognitive tautology interpreting thoughts through a modelled reality, modelled through thoughts through which they transform the unknown object of thoughts, a known theory applied to the reality they pretend to analyze.

Approaching the social reality through "*theoretical paradigms*"[114] does not raise any irritation in the heads of social science thinkers thinking about the social sciences, rather the opposite is the case: Presupposed thinking is considered not only a necessity of thinking, but the nature of creating—at least—social thought.

Epistemological specialists interpret the tautological operations of social science knowledge creation as a" dilemma" not to eliminate this dilemma, but to declare the dilemma as a necessity of thinking, preoccupied thinking as the nature of theorizing about the social.

> "Embodied in each research paradigm is a particular combination of ontological and epistemological assumptions, and these have a bearing on the kind of research outcomes."[115]

Social science thinking may heavily argue about different approaches, whether they are put forward by the schools of positivism, critical rationalism, hermeneutics or interpretivism, *not* making any determinating ex ante choice of any paradigmatic presuppositions, *not* making any determinating ex ante choices, already simply because one cannot know how to appropriately approach what is not yet known, is for social science thinkers thinking about social thought the only paradigm one cannot chose. Any thinking about the social must, according to the dogma of social science theorizing, construct social thought through such paradigms.

---

[114] N. Blakie, (2007) *Approaches to Social Enquiry*, 2nd Edition, Cambridge Polity Press, p 109
[115] Ibid

Chapter C: The social science approach to scientific thinking

Example 1:   The object of thinking, tuned to voice what social science thinkers want to hear from them

Thinking through a "paradigm", interpreting the object of thought adjusted to ex ante assumptions has developed a mode of thinking, in which the object of thinking is constructed for thinking, which consists of a monstrous apparatus of methods, Hegel's famous "sticks and spears", with which social science thinkers trim their thoughts to generate knowledge, which during the history of social sciences further developed towards a true technology of theorizing, called empirical research. Empirical research is a way of thinking that thinks about its presuppositions through a re-composed object of thinking, called the empirical reality, which is, different from the reality, the composed reality, composed for creating thoughts, knowing across all sorts of methods one "spear" that incorporates all the features of the social science procedures of thinking, the "indicator".

An indicator is the loudspeaker of the empirical reality tuned by the social science thinker to tell the thinker what the empirical reality should tell him. In his famous book *"Homo Academicus"* the famous sociologist Pierre Bourdieu discusses insights such as the following:

> "Indeed, rather than decipher one by one the different statistical relations, such as that which links the rate of divorce (an index of weak family integration) with having few children (a presumed index of weak family integration and above all of weak integration into the social order), we should try to grasp as a whole all the insights into social significance offered by the whole set of indices associated with the dominant pole of the academic field—large family, and Legion of Honour, right—wing voting and teaching of law...." [116]

What follows from this, kindly *"offered by the whole set of indices,"* is the whole set of common sense prejudices, in Bourdieu's study ennobled towards indices in empirical research about the homo academicus. Though, admittedly, Bourdieu's indices are documents of a true festival of tautological thinking (the rate of divorce indicates weak family integration and few children, weak integration into the social order) and of the academic regulatory authority sociological thinking has, it is much more important to notice that thinking in conditioning the reality for what the thinker wants to

---

[116]   P Bourdieu, (1988) *Homo Academicus*, Polity Press, Oxford, p 49

be told by his indices, a conditioning called the empirical reality, for social science thinking, making the reality say what they want to hear via the loudspeakers the thinker implants, is not an accident but the most noble way of social science theorizing. For social sciences, this is thinking about the real facts and not mere speculative theory.

After making the reader dizzy with his epistemological skepticism and with what this famous sociologist has learned about the many traps of theorizing about theorizers, in this study, not very modestly presented under the title "Homo Academicus", promising knowledge about the world of academics, he pulls the reader down to listen to the loudspeakers he without any skepticism implanted in the two faculties of the Sorbonne and bores the reader in more than 300 pages with what his loudspeakers told him about the "Homo Academicus", such as their age, sex, gender, "rate of divorce", number of children, large families and so on and on—in short all such data giving us a deep insight into the nature of what?—of the homo academicus.

Sentences such as,

> "Because of the fact that the accumulation of academic capital takes up time (which is evident from the fact that the capital held is closely linked with age), the distances, in this space, are measured in time, in temporal gaps, in age difference."[117]

This bacchanal of tautologies is not an accident but illustrates a masterpiece of social science thinking, a collection of common sense prejudices about such typical academic features such as the relations of age, sex, divorces, number of children, voting habits, political positions and so on and on, all building "social capital", a notion, which made this sociological guru famous and which this profound thinker certainly gained through pranced studies like this, studies he modestly calls a "prosopography of the university" [118], thoughts which have nothing much to do with what academics are at all.

And what do we learn about the homo academicus, lengthy presented on more than 300 pages, all stuffed with tables and data presenting such kind of findings about the "homo academicus",

---

[117] Ibid, p 87
[118] Ibid, p 39

one could just as well find about bus drivers or prison guards? We learn that the right wing, the conservative academics are those from the departments for jurisprudence, those who read—hard to believe but now once and forever proved by sound empirical thinkers—conservative newspapers, those who are rarely divorced and so on and on; left-wing people are, as is Bourdieu, often—sociologists. What a relieving insight for a left-wing sociologist to not be found as after all being those who are reading conservative newspapers. In short, we learn that all the stereotypes, not only all those left wing academics like Bourdieu shared before he published this book, are now "findings" proved by a deep sociological look into what the real facts are telling us.

A "'Book for Burning'?"[119] is the question he raises in his narcissistic introduction? A book that is saying anything about academics? Certainly not, but a brilliant example for studying the cognitive operations in social science thinking, certainly.

Example 2: The object of thinking, inviting social science thinkers to play with interpretations through theories they prefer

Common sense prejudices are only one way to construct cognitive models of the objects of thinking through which they, the objects of thinking, are asked to give the social science thinker their feedback on his hypothesis.

More common in social science thinking is to theorize on the objects of thinking through given theories:

> "One way to view American college life is through the theoretical paradigm of 'community of practice'."[120]

Thinking about the unknown object of thinking through already existing thoughts about it, does not raise the questions, why it is at all necessary to create knowledge about something if one already knows what it is or, if it is not known, how one can decide about a theory through which they operate thinking about the unknown.

---

[119] Ibid, p 1
[120] Charlebois, J. (2006) 'Community of Practice Involvement Obligations', *Journal of Intercultural Communication*, p 12

Approaching objects of thinking through prior chosen theories is in social science theorizing is not a theoretical mistake but a cognitive must: To subsume objects of thinking under the already existing knowledge thinkers own is an inevitable risk of thinking the more the thinkers knows. Social science thinkers appreciate the freedom of thought and define pre-occupied thinking as an epistemological necessity, "ordering schemes", a kind of thinking guide, without which no cognitive order of social thought could be created.

Not very much bothered by the question whether college life is such a community of practice to which this theory can be applied, nor if his interpretation of this paradigm coincides with the theory he chose, not to mention if this theoretical paradigm is a theory about such communities at all, this sociologist decided to think about American student's life by defining his ex-ante "ordering scheme" as a case of a "community of practice".

And what is a community of practice?

> "...As a result, much of our institutionalized teaching and training is perceived by would be learners as irrelevant, and most of us come out of this treatment, feeling that learning is boring and arduous, and that we are not really cut out for it. So, what if we adopted a different perspective, one that placed learning into the context of lived experience of participation in the world? What if we assumed that learning is as much a part of our human nature as eating or sleeping, that is both life-sustaining and inevitable, and that—given a chance—we are quite good at? And what if, in addition, we assume that learning is, in its essence, a fundamentally social phenomenon reflecting on our deeply nature of humans beings being capable of knowing? What kind of understanding would such a perspective yield on how learning takes place and what is required to support it? In this book I will try to support such a perspective."[121]

A "community of practice" is a theory about "institutionalized" learning, *"assuming that learning is as much a part of our human nature as eating or sleeping."* This social science thinker does very well know a distinction between *"institutionalized teaching and training"* and learning, but he decides to ignore the distinction he makes in order to identify learning as such with institutionalized learning just like eating and sleeping. This, ignoring the attribute—institutionalized—he attributes to learning, avoids raising the question why institutionalized learning *"is boring and arduous"*, because he is determined to create a theory about learning that pre-

---

[121] Wenger, E., (1989) *Communities of Practice, Learning, Meaning and Identity*, Cambridge University Press, p 3

sents learning "*as much a part of our human nature as eating or sleeping."*

Admittedly, one might raise the question if the theory about "communities of practice" is anything else but the determined wish of a pedagogue painting a rosy image of institutionalized learning. However, the way he constructs his theory about learning as a "community of practice" is exactly the same way as the Nobel Prize winner constructs his theory about the equitation of demand and supply. In both cases social science theorizing means to call the accordingly prepared social realty as a witness proving an ex-ante image they have about the object of thought, an "ordering scheme" through which they create their theories, theories "framing" their creation of theories discussed via an accordingly constructed "empirical reality" indicating proofs for their ideas, they always *"find"* after imposing them into the reality via their indicators.

Implanting a presupposed view on the social world into the social world via a theory through which the objects of thinking are "approached", imagining the social conditions of life the bourgeois society provides as a means for humans and if they do not, to then quite generously invent a theory that interprets them as a means for humans, here a theory about institutionalized learning, and to then "find" these presupposed views as what the "real facts" are telling the thinker, is the mode by which social sciences create their thought.

Example 3:   The object of thinking, prepared to prove what social science thinkers want to believe about it

Psychology has radicalized the idea to find the thought they want to find in the real facts and created "experiments" with humans to prove their pre-supposed theories.

Most appropriate for experiments proving a reign of mind instances behind human minds are procedures that extinguish the human will to prove or to use creatures without any will. Ducks, apes and dogs, animals are therefore populating theorizing about how humans will in psychology.

In order to measure what all their theories say, that there must be the ability towards an ability, theories about intelligence argue for the ability to think, intelligence, and construct their proofs of the ability of intelligence through the absence of knowledge to prove the presence or absence of intelligence.

The pretended similarities of experimental thinking in psychology and natural sciences are however the sheer opposite: While the natural science generate through experiments the interrelations of any elements they want to think about and therefore exclude interfering impacts, psychological theorizing excludes what it intends to prove as being absent to prove the presence of what their theories state.

Example 4:   The object of thinking, offering social science thinkers to ennoble political standpoints as scientific insights

The real art of social science thinking is to present the re-discovery of the presuppositions through which the object of thinking is approached as a conclusion from the composition of the objects of thought, through hypothetical thinking as a conclusion from its observations of the data, conclusions the thinkers—as they phrase it—ever "find" in the social reality.

By no means, this is the exclusive art of "Western" social science thinking, practiced in the "Global North", but this is the mode of thinking, "Western" sciences succeeded to universalize across the world as the universalized mode of social science theorizing.

> "If we consider the parallels between economic dependence and academic dependence we may define the latter as a condition in which the social sciences of certain countries are conditioned by the development and growth of the social sciences of other countries to which the former is subjugated."[122]

This famous Latin American critic of the global social science world presents his conclusion not even as a more or less disguised tautology, but as a mere repetition of his presupposition: Yes, if we consider academic dependence as the same as economic dependence then academic dependence is the same as economic dependence. The only thing that this statement presents as if it was an insight discloses only that it is the determined will of this thinkers, to see the world of sciences as he is determined to see it. This insight is

---

[122] Dos Santos, Theotonio, (1970), *The Structure of Dependence,* American Economic Review 60

not "excellent", but it is honestly presenting his insights as his mere decision to see things as he decided he wants to see them.

However, with or without any pretended conclusions, his view is though only false in many senses. Considering the exploitation of developing countries as "*economic dependence*"is not only an euphemism, that does not want to see that these countries are not just a disadvantaged players in the global economies, but a mere resource for exploiting their resources, an exploitation in which the exploiter plays the role of both the buyer and the seller of the resources these countries only in a very formal sense own and sell.

Calling such relations "economic dependence" and applying this false judgement about the economic relations to the world of sciences as "academic dependence" only reveals the political will, that this thinker does not want to critique, interpreting social thought a matter of a global knowledge market and academics the intellectual troops of a competition among nation states over the resource knowledge, but that he wants to critique the "unequal" conditions for these national battles about knowledge. In this sense, his presupposition to apply the notion of economic dependence to thinking about the world of social thought has allowed him to articulate his political presupposition, full of euphemistic interpretations about the economic relations between developing countries and their exploiters and full of affirmative views about the global sciences, presented as if it was concluded from the science world.

In doing this, presenting his political convictions as a thoughtful conclusion from his observations of the science world, he finally also proves, how independently this thinker imitates the techniques of social science thinking and thus also proves that the science policies of nation states in "developing countries" are no doubt in a truly disadvantaged position in their competitions with the "North"—but only because they are so keen to join this competition, they could also oppose or at least not join. Then, they could just critique social thought coming from "the North" instead of imitating social science theorizing as the arrangements of theorizing as a completion among theories to then complain that their own theories are not competitive. They—unfortunately—are, one must say, considering their successful imitations of the mode of social science thinking, however, as in many other cases of imitations, the originators of social science thinking are often just better in presenting their determinative false thoughts as if they were any insights.

Why teleological thinking must be the nature of thinking

Social science thinking is not only coincidentally, by mistake preoccupied, teleological thinking. Social science thinkers thinking about the social sciences heavily confirm that preoccupied thinking is the nature, not of social science thinking but of any thinking.

The epistemological expert department of social science thinking, which engages their most sophisticated social science thinkers to practice social science thinking about social thought, know why non-determinative thinking is not possible and that thinking must ever fail, may this be as "trial and error" or discussions like those all the Luhmann's appreciate to discuss, raising such enlightening questions as this one: "Are social sciences possible". The latter one is a question most distinguished scholars do not raise before doing social science theorizing to decide if they should practice science or not, but they raise the question, if social sciences are doable to practice what they doubt can be done as bearing witness to an exquisite level of reflexivity. And it is this discussion of this question *which is already the whole insight in science* they create about social science theorizing, however, it is not at all only this more recent variation of practicing science as doubting if social sciences are doable.

Social science thinkers thinking about social sciences in all the previous cases of an epistemological skepticism never became schizophrenic about practicing science as doubting science, doubting if what they do is doable, as they advocate this very schizophrenic concept of scientific thinking questioning its possibility as a necessity of any theorizing, and the meta—debates in social sciences do this since the existence of social sciences as *the* essential of reflecting about social sciences. It is indeed this one and only conviction, to ever scientifically prove that scientific thinking must ever fail, that founds all the epistemological considerations about social sciences.

The logic in all their various attempts throughout the history of epistemological thinking in the social sciences is, since Kant and his "*Kritik der reinen Vernunft*", always the same: social science thinking must fail, because, no doubt, thinking *can* fail. It needs however a very determined will of thinkers to mistrust the human ability to think, to conclude, just while saying that they *know* how to distinguish false from right knowledge, to conclude from this observation about knowledge that knowledge can be wrong or right that therefore knowledge can never be right.

As for example the founder of positivism knows:

> "Because the social scientist is a member of the category of the phenomenon being studied, disinterested detachment may not be possible."[123]

And what if the thinker—who obviously not only knows that being detached to practical interests violates thinking—what if he uses his brain, uses the knowledge he has about being a "member of the category of the phenomenon being studied" the thinker obviously owns, to distance himself from being a member of any "category" while he is thinking and avoids any determining choices, choices he only he makes, especially since he knows what his preoccupied choices are and even, thanks to the epistemological gurus also knows how preoccupation choices undermine the creation of knowledge? No, this, excluding any preoccupations the thinkers knows from thinking, is impossible, thinking cannot escape from the imprisonment into which thinkers put themselves despite their knowledge about the cognitive damage detached thinking causes for theorizing and therefore must be teleological thinking—with one exception: Only profound epistemological thinkers, members of the epistemological category of thinkers passionately denying the possibility of disinterested thinking, are able to escape from their own epistemological prison and tell us the very final truth, if not about any social phenomena, but at least about scientific thinking as a whole and they do this since the inception of social science, insisting that social thought must be presupposed thought, of course with the exception of this objective, non-presupposed dogma. It is hard to imagine how one could better justify the alignment of the thinker with his object of thinking as presenting the necessity that thinking must ever fail as a necessity of the object of thinking. Since the scientist is a "member" of the social he thinks about, his thinking must share the objectives of the social he thinks about

---

[123] N. Blakie, (2007) *Approaches...*, p 34

and imposes them into his thinking through the "paradigms" through which he constructs his thought ever "finding" in his object of thought what he must find thanks to his epistemologically inevitable detachment with his object of thinking.

According to social science theories about social science knowledge any knowledge must be preoccupied knowledge. The common sense dictionary, Wikipedia, confirms the scientific common sense according to which thoughts, constructed from their modelled pre-suppositions are proved against their coincidence with the social reality.

The concept of "evidence" in fact accumulates all the false theories social sciences have about scientific thinking and this concept confirms the affirmative nature of social science thinking they proclaim as the nature of social thought.

> "Evidence is information, such as facts, coupled with principles of inference (the act or process of deriving a conclusion), that make information relevant to the support or negation of a hypothesis."[124]

In this concept of proving thoughts via their evidence, thoughts must be proved by re-migrating to the phenomenological appearance of the object of thoughts thinking intends to overcome when mindfully going behind the surface of things, the wandering between the phenomenology of the objects of thoughts and the essence they do not reveal to us without thinking, which illustrates in this way of proving thoughts the nature of the above duplication of the object of thinking and into an object of thought and an instance judging about them.

The concept of evidence, that thoughts must be proved against the phenomenology of the objects of thoughts, is not only a cognitive contradiction, sending the mind penetrating the phenomena to reveal their nature, to detect the essential social objectives hidden behind their reified appearance, back to their reified appearance to prove their thoughts. This, sending thoughts back to the reified appearance of the object of thought, subjectivating them under the reign of facts and, thus, under the reified objectives incorporated in the object of thoughts, travelling back and forth between descriptive thoughts and the phenomenological surface of phenomena, never digging into the essentials of any social constructs, ever

---

[124] https://en.wikipedia.org/wiki/Evidence

comparing the disconnected descriptive observations with the never understood social objectives behind the disconnected phenomenology of facts, is the cognitive birth of the methodological affirmative nature of social science thinking.

Indeed, scientific thinking that practices the dogma of the claimed necessity of presupposed thinking, not coincidentally, but systematically creates thoughts through presupposed thought, is the social sciences way of thinking and, claiming that any knowledge must be wrong, does neither make this claim nor all the wrong knowledge right, nor does it make any knowledge right, that beforehand confesses its limitations.

Hence, since then any abstruse pre-assumptive thoughts are ennobled as science by the nobles of science and, as in our first case, rewarded with a Nobel Prize.

However, exactly confessing this, is the aim the epistemological dogma is aiming at. Disarming social thought from its only weapon, its knowledge, is the political aim and motivation, why the nation state society's epistemological science police so passionately argues against knowledge. Downplaying science towards a mere opinion reveals this utterly political position about scientific knowledge as the cognitive ethos behind this dogma.

Epistemological social science thinkers are, no doubt, consequent in the construction of their spiral of false tautologies: If thinking is to re-detect what the thinker has imposed into the object of thoughts via his presupposing way of thinking, then, voila, *"disinterested detachment may not be possible."* Hence, just as social science thinkers re-detect in their objects of social thought what there presupposing ex ante thoughts have imposed into them, epistemological thinking re-detects their epistemological presuppositions in their theories about theorizing: Presupposed thinking is the nature of social thought as a social science guru thinker such as Habermas confesses:

> "Theories are ordering schemes, we discretionary construct within a syntactical binding framework."[125]

At least, as one can see from the epistemological supervisors of social science thinking, my above example chosen from an economist

---

[125] J., Habermas, (1982) *Zur Logik der Sozialwissenschaften*, Frankfurt a.M., S. 17

was not the pre-occupied selection, only an interested choice to show the technics how social science thinking approaches its objects of thinking, not an exceptional way of theorizing in this example from economic theory, but, proved by their epistemological druids as an example that represents not only the social science way of thinking, but, according to the social science epistemological gurus, the nature of any scientific thinking, thus complementing the very Christian images of the disciplines about the "unfinished human" with a Christian image[126] about the limits of the human intellect limited to the moral imperative of the relativism of a democratic opinion.[127][128]

The stigma of the natural sciences—and the self-destruction of an envied hero

Presupposed thinking must not only be the nature of social science thinking, but the nature of scientific thinking as such. Thus, the natural science in which presupposed thinking is a violation of scientific thinking and, unlike in the social sciences, not practiced as a cognitive must, thus creating objective knowledge and no pluralism of presupposed knowledges, natural science thinking gets into the sight of social sciences reflecting about scientific thinking. Natural sciences violate the theories of social sciences about social sciences

---

[126] Unlike this very Christian image of science demonstrating with its scepticism the limits of science, the concept "ilm" in the Islamic social sciences considers knowledge as a means for the concerns of social practices: "'Knowledge' falls short of expressing all the aspects of 'ilm'. ...Ilm is an all embracing term covering theory, action and education." S.W. Ahktar, (n.d.), *The Islam concept of Science, part 1,* Al-Tawid, A Journal of Islam Thought and Culture, 12. No 3

[127] The inevitable question of whether what has been said in this chapter about the particular way of social science thinking can be "generalized" across all social sciences, I leave to the social science thinkers who do not want to understand that counting the number of the same phenomena and that raising the question which of the non-understood phenomena can be subsumed under a non-understood category, is the opposite of thinking. The more of the same one can find, does not make it more comprehensible.

[128] To avoid again a misunderstanding: This critique of the social science way of thinking by no means replaces any critique of any theory this way of thinking creates. This, the critique of theories created through this approach to science, is the hard intellectual work that is due to be done with these theories and with any other theories created with any other approach to thinking.

by violating their one and only epistemological dogma that knowledge must be presupposed knowledge.

Before discussing the theories in the social sciences about natural sciences, it should be clarified that reflections on the natural sciences, which do not focus on their knowledge, but on their ways of theorizing—are no natural science thought but social thought.

The debates among social sciences, may they be carried out by social or natural scientists are characterized by an odd ambiguity: Social sciences discuss natural science both as a positive example of "hard" science, envying them for creating knowledge and as their opponent, since the very knowledge they create disproves the social science theories about the necessity of a knowledge relativism. Thanks to the help of a natural scientists shifting to the métier of thinking about the social, epistemological thought about social and natural sciences, social sciences managed to merge both views in the same judgment about the natural sciences.

The envied hero....

Only social science thinkers, who rightly critique the speculative nature of the classical philosophies and who from there make the wrong conclusion to enthrone "the real facts" as a cognitive means of thinking, only thinkers for whom the journey between their presupposed "hypothesis" and the phenomenology of the objects of thinking, are caught in the belief that thoughts must be "found" and proved via their "evidence" shown by the phenomena they think about. Using the lacking knowledge about the unknown object of thinking as a cognitive vehicle for creating knowledge is the bizarre idea in the social sciences about how to prove thought that founds the nature of social science thinking.

It is, indeed, *the* epistemological life lie of social sciences theories about social science theorizing, that social sciences prove their theories by comparing them with the object of thinking. Thought can never be proved or disproved with the object of thinking, but only with the observations the mind makes about any object of thinking and it is *the* epistemological self—betrayal of social sciences, making themselves believe that they prove their thought via comparing them with the object of thinking—just as the concept of "evidence" insinuates.

What they really do compare is their thought with their presupposed theories, through which they model the object of thinking into what they call the empirical reality and by doing this, the concept of evidence proves the affirmative way of social science thinking, using their theoretical constructs they construct through any presupposed theories to prove these constructs with what they have imposed into the object of thinking, the notion of evidence interprets as being voiced by the object of thinking.

As if thinking was not entirely superfluous, it is this theoretical nuisance, attributing—thanks to a deep skepticism in thinking—the object of thinking the cognitive ability to prove thought, that not only characterizes the affirmative nature of social science thinking, a way of thinking social sciences ennobled and developed with armies of thinkers towards their monstrous apparatus of thinking methodologies[129], it is a nuisance, armies of social science put into

---

[129] It might be this monstrous methodological apparatus, and the false believe that this apparatus coincides with the nature of thinking in the natural sciences, that social sciences gained the myth of being rational. Or is it simply due to the fact that the imperial nation states successfully execute their rationale inside their societies and outside against other nation states that creates the myth of rational social sciences in these countries? Anyway, it is this myth that occupies major debates of a false opposition against the social sciences, namely among scholars in the "developing world", ending up in advocating "irrationalism" as an opposition against social sciences, and thus even eliminating the false impression of rational thought social sciences create with their methodological apparatus. Nothing proves better the irrationalism of social sciences thinking but their false identification of causes in the natural sciences with reasons in the social sciences as exemplified in the false interpretation role of the experiment in the natural sciences. What troubles social sciences is what constitutes them, the human will, constituting reason, they firstly mystify into nature, causes, they wish to domesticate in the relation of effects and causes. The free will is the devil for social sciences, since it is the devil nation states must domesticate. And one needs to have a remarkable amount of a combination of hypocrisy, cynicism and ignorance to provide arguments, proving the theory that Africans are unable to argue with arguments and to thus most rationally, though falsely prove, that, after the imperial world has destroyed the traditional organisations of life of people in Africa during colonialism and then, once they became "independent" nation state, after some local elites educated in the imperial world, found under the guidance of the imperial world that they should imitate the organisations of live of the imperial countries of a nation state and a capitalist economy, the very imperial countries' economic exploitation and political interventions at the same prevented to establish, to consider the disastrous living conditions of African people in a twofold destroyed life organisation as the result of their

practice in their affirmative way of theorizing, that also constitutes all the debates of the nuisance of a mindful object of thought, when they discuss the natural sciences, repeating and reproducing this false theory about a thoughtful object of thought, namely by their false theories about the experiment in natural sciences.

Epistemological social sciences theories interpret the experiment in the natural sciences as the very cognitive means to prove thought, an ability they attribute to the object of thinking, and this is a misunderstanding of what the experiment is, as if the experiment in the process of natural sciences knowledge generation was the materialization of very cognitive means social sciences believe were the mystic abilities the objects of thought own to prove or disprove knowledge. This is an error, and this a pretentious error only thinkers can make who attribute the object of thinking an ability to judge about thought and it is mainly this error that is the bases of all the false debates among social sciences about natural science, including the contributions from natural sciences, when natural scientists discuss the nature of theorizing in the natural sciences.

Only such thinkers, attributing the object of thinking the ability to think about thoughts, to respond to thought the thinkers created, can articulate in their epistemological departments since the inception of social science thinking their jealous dream, social thought may be as objective as the thought of natural sciences, an objectivity they believe natural sciences thought gain by proving their hypothetical thoughts via experimenting with facts. This is however a quite intended misunderstanding about natural science thought and about the role of the experiments in particular, only thinkers can create, who idolize reality as a means of cognition and therefor misunderstand the experiment in natural sciences thinking.

Experiments, the arrangement of a certain constellation of natural phenomena do not prove insights, but are a construction of the objects of thinking, a construction the coincidental nature of nature does not provide, in order to find out insights about causal relations between elements of the nature, that do not exist in the

---

"irrational" nature, unable to build nation states. Thinking such thought is classical social science logic, where any makings of human life, however it may oppose each other, is ever the result of a human nature, a nature which always precisely coincides with whatever humans made from live throughout history. And this is rationale?

coincidental constellations of nature without the arrangements of an experiment. The experiment arranges the object of thinking and is no natural sciences methodology for proving theories and does not prove any thought. It is not a proof for any hypothesis, but the creation of conditions under which relations between elements can be observed, conditions which the nature does not provide. The experiment constructs the objects it wants to theorize about, relations between natural elements, cleared from others, which would not occur in nature.

The so called "Newtonian thinking" is a fiction of the epistemological gurus of the social sciences, misinterpreting—on the basis of the false interpretation of the role of experiments in the natural sciences as a method proving insights—that natural sciences invented a particular way of thinking, according to which they do, what the epistemological departments of the social sciences see as the way of gaining scientific insights, the rationale of social science theories about theorizing they incorporate in the objects of thinking as the way they scrutinize them. For social sciences the experiment is the realization of the cognitive means they attach to the object of thinking as its ability to prove or disprove knowledge.

Calling this, detecting the inner logic that constructs objects of thinking in the natural sciences as what they are, the causal relations found in the object of natural sciences thinking, making them why they are what they are, calling what natural sciences find in the objects of thinking, causality, the objectives natural science *thinking* obeys, and interpreting this subordination under the nature of the objects of thinking in natural sciences a particular way of theorizing,—falsely—interprets natural sciences thinking as the "Newtonian rationalism" and alike, and thus prepares to question this thinking that results in objective knowledge as only a particular way of natural sciences theorizing. It is this interpretation of theorizing in the natural sciences as "Newtonian thinking" as an episode in the natural sciences thinking that prepares the aim to prove the impossibility of objective knowledge in the envied natural sciences, the destruction of the scientific hero, only the social sciences constructed with their false theories about the natural sciences, namely the experiment's metaphysic cognition abilities.

...and his self-destruction

Insights that natural sciences shift their theories over history, from time to time critique their insights and arrive at new insights about the world—only proves the impossibility of thinking for those, who are obsessed with even interpreting the progress of knowledge as a proof of its impossibility. The consequence of the critique of the idolization of speculative thinking is not only the idolization of the reality as a means of thinking, witnessed by a false theory about natural sciences, it is the first preparatory step, to prove the impossibility of any objective thought, also in the natural sciences. Proving after all this, the impossibility of scientific thought, is the epistemological motivation behind all social thinking about the natural sciences, including thought of natural scientists, when they epistemologically argue about science, the natural sciences.

It was in fact the knowledge, created by the natural sciences that contradicted the social science theories about science, at least it questioned their view on science so far limited to the social sciences, until a natural scientists assisted social science thinkers to omit this stigma from social science theories about scientific thinking that only they are unable to create knowledge and questioned the insightfulness of the theories of natural sciences—according to social science thinking about natural sciences. Natural sciences are not affected by this debate and continue to create knowledge, also natural scientists question as being knowledge only once they theorize about science and apply the modes of social science thinking to think about natural sciences.

To finally extinguish the remaining rest of knowledge in the positivist view implied in the idea of a falsification of knowledge, a view which still contains in its concept of an ephemeral objectivity of knowledge a reminder of what knowledge is aiming at, it was the work of Thomas Kuhn on the knowledge progress in the natural sciences that not only finally applied the view of social sciences about the inevitable relativism of knowledge to the natural sciences and their knowledge, but also paved the way towards a radicalization of a relativism of knowledge in all sciences, later proclaimed by "poststructuralist" thinkers such as Levi Strauss and Foucault, advocating as the objective of knowledge, a "bricolage"[130] of partic-

---

[130] Lévi-Strauss, (1962) *La Pensée sauvage*. Paris, Éditions Plon, p 26–28

ularisms, "wild thinking" that elevates the psychological problems only competitive subjects have in social science psychological theorizing, their struggles with their identity towards an epistemological necessity of scientific theorizing, science as a "whateverism" in which the absurdity of arbitrary theorizing is proclaimed as the objective nature and aim of scientific thinking.

In this book T. Kuhn describes numerous examples of how in the natural science false theories are corrected. The notion of a shift of paradigm is what has made this book famous and is a wild nonsense, as Kuhn himself later, but too late, realized. Concluding from Copernicus' corrections of a false theory about the relation between sun and earth, which was exploited by all kinds of mystic philosophical and religious debates about values, beliefs, concluding that it is this shift of moral values, T. Kuhn calls paradigms, a kind of mysterious cloud of moral thinking, a *Weltanschauung*, which is the reason for Copernicus' theory shift, and thanks to the interpretation of this notion of a "paradigm" that such mystic mind settings are incommensurable, meaning that they are beyond the reflections of sciences, is a construct of a mystic notion to explain the progress of knowledge that is very typical for social science thinking about science.

Only because it—naturally—takes some back and forth in thinking and to finally detect a new theory as a the correction of a previous theory, only because it might take some time and goes through theoretical battles, one does not need to make a mysterious question from the fact that from the point of view of the new knowledge, that now, after false theories have been detected as false theories and have been eliminated from the knowledge as false knowledge and after the new insights have been shared by everybody as correct knowledge, making ex post a mystery from this progress of knowledge and to argue that it is hard to understand why it is so difficult to overcome a knowledge that was seen as correct knowledge, only mystifies the progress of knowledge—in order to create a very false question. Interpreting the most detailed descriptions of how knowledge in the natural sciences progresses or fails to progress, described by Kuhn along all kind of case throughout the history of social sciences, concluding from these insights into how the knowledge in each case progressed or not, raising from such insights explaining how knowledge progressed or hampered to progress the question, why it progressed and why and where not, raising this question why it progressed after all has been said about how it progressed, remains only an unresolved question if one is determined to create a mystery about the time and circumstances it takes to shift towards amended theories. Only once sharing the progressed knowledge and finding it so convincing compared to the previous false theories and compared to the insightfulness of the critique of the faults of an overcome theory, one can raise the false question, how come that thinkers could stick to any old false knowledge to then conclude from this mystification of the why they very well know, that there must be anything mysterious beyond the consistency of theories that decides about the progress of knowledge. Only because a new theory is considered now as so insightful, one can construct the question, if people did not right away share this insightful theory, that there must be anything beyond the insightfulness of the thought of a theory, from which the insightfulness of theories, their progress of knowledge is dependent.

For social science thinking Kuhn's ideas about "paradigms shifts" that are needed to shift from a false to a corrected theory, demonstrate—for social sciences—what social sciences always say about how thinking in the social sciences works and what it creates: Presupposed thinking resulting in relativated knowledge, now also applies to the natural sciences—just as if Copernicus' new theories were driven by any ex ante *Weltanschauung* calling for the view that it was the earth revolving around the sun and, only as a consequence of this ex ante moral *Weltanschauung*, Copernicus managed to critique the existing theories and to provide his new insights as a response to the needs of this new *Weltanschauung*, thus also only a relative insight.

The idea of a paradigm that is made responsible for correcting false knowledge, affirms the convictions of preoccupied thinking of epistemological social science thought that if the history of natural sciences is a history of correcting false theories, that this then proves that—so the pretentiously false conclusion,—even—in the admired and envied natural science objective knowledge is also ever relative knowledge. May T. Kuhn be famous for ever for his theory, but his conclusion from his false observations of a mystery of shifting paradigms guiding the generation of natural science knowledge and their progress of knowledge, is as false as a conclusion can be. It is as if one concludes from the fact that anybody who does not drown must be a bad swimmer.

If one forgets about Kuhn's mystic construct of a paradigm (which in his later work he is actually not only no longer using, but which he questions as articulating what he really wanted to say in his book that made him famous), it is true that the history of natural sciences is a history of correcting false theories and of closing theory gaps. What else—knowledge progresses and the progress of knowledge is about omitting false theories—what else! However, concluding from the progress from false to correct knowledge and from closing knowledge gaps, that knowledge must be ever false knowledge, only makes sense for somebody who is obsessed with the idea of the relativism of science and, who is not only therefore mad enough, to insist on the dogma that there is no correct knowledge as a correct thought about thinking, but who proves this dogma that there is no right thought with the corrections of false thought. Why then not correct this obvious false conclusion? Because it would need a correction of the dogma, a dogma which really deserves to be mystified as a "paradigm", a dogma which makes social sciences what they are, thinking that is and must be preoccupied thinking and therefore must be ever relative thought.

Kuhn's book, which discusses the knowledge progress in the natural sciences, is so appreciated by the social sciences, because it answered a false epistemological question with the accordingly consequent and therefore false answer, and by doing this, abolished the major pain of social science thinking about the natural sciences—which is that his books proves the only objective insight social sciences insist on, proving the impossibility of objective knowledge, now also in the natural sciences. It was a pain for social sciences epistemologies, that the knowledge of natural sciences disproved this fundamental dogma of the social sciences with their objective theories and with their real progress from false to right knowledge, a progress that is essentially impossible in the scientific relativism of presupposed theorizing. It is the contribution of T. Kuhn, thinking as a social scientist on natural sciences, to confirm with social thought of a natural scientist about natural sciences, that it is a false theory to insist on the objectivity knowledge in the natural sciences and that natural science knowledge is also relative, thanks to the striking logic, that natural sciences correct their mistaken thought, and thus to prove, that the social science dogma, that there is no objective knowledge, also applies to the natural sciences, thus finally proving the absurdity that it is a correct thought that there is no correct thought in no science.

The decline of scientific knowledge towards ephemeral knowledge

The destruction of the natural sciences knowledge towards also relative knowledge via the invention of a mysterious paradigm shift, implies an assumption, another false theory, which is to identify the distinction between false and right knowledge with the *acknowledgement* of knowledge as false and right knowledge. The question Kuhn is really raising is not how knowledge progresses, this question he answered along all his examples. The question he is really raising is the question, why and how new knowledge becomes acknowledged knowledge or shared knowledge.

Concluding from the progress of knowledge in the natural sciences towards the relativism of knowledge in the natural sciences, is not only a false conclusions, the question Kuhn raises is a question he only raises thanks to a false concept of knowledge, more precisely a concept of true knowledge, a concept only the social sciences know, which is to identify the correctness of a theory with a theory that is shared or in the case of a false theory a theory that is not shared and, hence, only arising from this false identification Kuhn seeks to answer the question what is needed to shift from one shared knowledge to another shared knowledge and to arrive at shared new knowledge. Unsatisfied with all the answers he very well gives in all his examples about how in each case knowledge progressed or not, only because he identifies right knowledge with shared knowledge he feels puzzled to raise and answer another, now false question, which is to answer the question how the sharing of knowledge shifts from old to new knowledge, a question into which he falsely translated his question about how knowledge progresses.

His answer is on this false question is a false as it inevitably must be, it is that the progress of knowledge only progresses, once a new "paradigm" is a shared "paradigm".

"A paradigm is what the members of a scientific community, and they alone, share."¹³¹

Explaining a shared "paradigm" as the precondition for sharing new knowledge is an obvious tautology T. Kuhn does not realize as this tautology. The reason for this is that he considers the explanation of the progress of knowledge as the same as shifting from one shared knowledge to another shared new knowledge, which inevitably result is the tautology of his paradigm shift as a condition of a paradigm shift, the pleonasms of a shared paradigm, a tautology which is the inevitable consequence of reflecting on the progress of knowledge on the basis of identifying right or false knowledge with non—shared or shared knowledge, a theory he borrows from epistemological reflections in social sciences about social sciences he applies to thinking about the natural sciences and how their knowledge progresses.

And this theory is a false theory about knowledge, and applied to reflect on the progress of knowledge inevitably ends up in the tautology of shared knowledge as the condition of sharing knowledge, Kuhn disguises in his mystic tautology of a "paradigm" shift towards a "shared paradigm".

To seek the answer for his false question about why knowledge progresses and finding his answer in his mystic "paradigm" T. Kuhn indeed interprets the distinction between false and right knowledge as the same as acknowledged or non-acknowledged knowledge, identifies the scientific quality of a theory with its perception and therefore also seeks beyond the contents of theories for any contexts to answer his question under which circumstances knowledge is acknowledged as a correct theory, in order to prove that—as if it was an irony—just when knowledge overcomes false theories that correct theories are—in this historical sense—ever only relatively correct theories and from this historical view on the progress of knowledge concludes towards a relative nature of knowledge.

It is not that important if it was T. Kuhn himself or all his passionate followers[132] within the social sciences who finally made

---

[131] Kuhn, T., *Second Thoughts on Paradigm*, http://eu.pravo.hr/_download/repository/Second_Thoughts_on_Paradigms.pdf
[132] To quote just two examples. The first one from the above essay in his 50th anniversary book: "Kuhn does not permit truth to be a criterion of scientific

what his work stands for among the social sciences and his later self-criticism about the notion of a paradigm, might be a hint that it was rather his readers from the social sciences who made from his work what they read into it.

In any case, it this book that raises this wrong question only epistemological thinkers in the social sciences can raise: This wrong question is, how science progresses and how scientific revolutions occur, which is for him the same as asking how scientific theorizing acknowledges a theory as a valid theory. And this question was and is a wrong question, since it part tout does not want to find the reasons for the progress of knowledge in the progress of knowledge, the progress theories made, rejecting an insight as false and shifting to new insights omitting mistakes, corrections of false thought Kuhn describes and analyses in his book in such details along so many examples and at the same time asks the false question why these corrections could be not made at any other time under any circumstances needed to correct false thought and thus seeks and finds an answer on this false question in the dragging acknowledgement of corrected theories as correct theories.

Put in other words, his false question is to ask, why it happens that a theory is not acknowledged as a theory, *though* it exists as a correct theory and as a consequence seeks for an inevitable mystic something that makes thinkers *acknowledge* a theory, a question which again inevitable ends up in the poor tautological answer, because they did not share the same knowledge, a tautology hidden behind the mystic notion of a paradigm: Knowledge shifts because it shifted. "Paradigms" most dubiously circumscribe tautologically the circumstances, when increasingly more thinkers acknowledge a new theories, because more thinkers acknowledge a new theory and by doing this, creating the paradigmatic circumstances for a shift of paradigms.

---

theories, he would presumably not claim his own theory to be true." https://books.google.it/books?id=xnjS4o1VuFMC&dq=thomas+kuhn&hl=en &sa=X&ei=dy6gVIOYIoOvU_nEgMgL&redir_esc=y
The second one from the US philosopher Thomas Nickels: "Clearly, paradigm change is not a rational process as understood by the traditional canons of rationality....Yet, in Kuhn's own view paradigm decisions need to be irrational." Nickels, T. (2003) *Introduction, "Thomas Kuhn, Contemporary Philosophy in Focus"*, Cambridge University Press, p 2

To make false or right knowledge the same as shared or non-shared knowledge is a false theory about science, more precisely about the concept of "truth" that identifies true knowledge with shared knowledge. Identifying both, right and shared knowledge is the concept of a *discursive 'truth'*, is *the* mistake that originates from the social sciences dogma that knowledge must be preoccupied knowledge and applying the concept of a discursive truth to the question how knowledge progresses, results in the very tautological answer on this question, revealing the nature of scientific discourses only the social sciences discourse practices, handling the contradiction of their dogma, creating ever only relative objective knowledge.

Social sciences insist on the contradiction of a relativism of knowledge, on knowledge that must be proved as knowledge though must ever be relative knowledge and therefore export the criteria against which they measure what knowledge is knowledge and what not from knowledge into the discourse about knowledge, resulting in the odd construct that it is the subjective agreement among scientists to identify relative knowledge as objective knowledge, once it is shared knowledge.

Theories may be found wrong or right by any receptors of theories. If one, however, considers wrong and right theories as the same as what their receptors think is wrong and right knowledge, than this is a false thought. It is a false thought, since the question, if a theory is false or right does not depend on the ability or willingness of a recipient of a theory, but on the consistency of its proves. If it was the case, that theories are wrong or right dependent on the ability or willingness of the recipients of theories, theories were wrong or right because of the abilities or willingness to share the thought of a theory or because recipients are not able to understand or follow them. Copernicus' theory was not false because recipients did not share it. The lacking insightfulness to distinguish right from wrong thought in a theory is rather an argument against the ability of the recipient, never against the theory. Or to put in other words, the question what finally is a wrong or a right theory in no case can be decided via the question if a theory is shared or not.

It is this false question and its false answers that invites all the social sciences celebrating this work as a milestone for how to see the knowledge of the natural sciences. This question is false, since it concludes from the fact that correct knowledge exists, but that is not without any disruptions acknowledged as correct knowledge, that one must therefore look at any circumstances outside of the knowledge to answer the question under which circumstances humans change theories. This is a false question, since it seeks the reasons for the quality of a theory beyond the theory in the cognitive abilities or willingness of its recipients to share a theory or not. If this was the case, the theory which saw the earth as a disk, was a correct theory until Copernikus. Or: If this was the case and given the catholic church would still rule thinking about the world and not the modern society with their capitalist and their interests in natural science knowledge, the earth would still be a disk, and this would therefore still be a false theory, though it would be not an acknowledged theory. In short: If wrong or right knowledge was a matter of which knowledge is acknowledged knowledge all the false knowledge the progress of knowledge in the natural sciences identified as false knowledge was correct knowledge until it was corrected. And if this was the case, if knowledge was knowledge was dependent on the question if it is shared knowledge, then both false and right knowledge could be shared knowledge, hence, if knowledge is knowledge or not is decided by the question if it is shared knowledge or not, one could not even distinguish between false and wrong knowledge, since both are—historically—shared, thus right knowledge, in other words, there was no way how it then could be possible to at all distinguish false and right knowledge, not to mention to think about any progress of knowledge, about any "scientific revolution".

With this in mind one can better understand, why T. Kuhn's book, possibly more precisely, how his book was mainly interpreted by the social sciences became an important piece of science and how this debate, his book inspired, how this debate made a such an essential contribution to the dissolution of scientific knowledge towards a mere opinion, towards true knowledge as ephemerally true knowledge.

It is a matter for biographers, if it was after all T. Kuhn, himself or if it was the reception of his book by the social sciences, in any case, surely explaining "scientific revolutions", the progress of knowledge as a shift of paradigms as Kuhn did at least invites all the interpretations of his work, that made his work famous, more among social sciences than among natural sciences his book is though about—and a matter of the social science criticism.

It is this criticism that is very informative about what made his book so much appreciated by social sciences.

> "And so Kuhn was accused in some quarters of denying the very rationality of science. In other quarters he was hailed as the prophet of the new relativism. Both thoughts are absurd."[133]

Thinkers like T. Kuhn, as he did, may very well deny, being interpreted as relativists, but it is this identification of the scientific substance of a theory with a theory that is shared, that invites relativists like Popper and his successors, to consider his work as proving the relativism of knowledge—as much as Kuhn may ever deny that this was ever his intention[134]

> "Granting that neither a theory of historical pair is true, they nonetheless seek a sense in which the latter is a better approximation to the truth. I believe nothing of that sort can be found."[135]

The fact that T. Kuhn even distances himself from the relativism of an ever approximative knowledge, does in fact not mean that he rejects his own relativism of a historically 'true' knowledge that considers knowledge as 'true' knowledge, as long it was historically acknowledged as knowledge. He therefore does not want to see any corrected knowledge as simply false knowledge. It is this historical relativism that inevitably and rightly feeds his theory about the progress of knowledge in the natural sciences as a theory that proves the relativism of the natural sciences knowledge.

Only a thinker, who approaches the most detailed description of the knowledge progress natural sciences—a progress the natural sciences unlike social sciences, as T. Kuhn very well knows—do

---

[133] Hacking I., *(2012) Introductory essay,* in: Kuhn, T., *The structure of scientific revolutions,* 50th anniversary edition, The university of Chicago Press p xxxi

[134] "Granting that neither a theory of historical pair is true, they nonetheless seek a sense in which the later is a better approximation to the truth. I believe nothing of that sort can be found." Kuhn, T. (1970) *Reflexions on my Critics,* in: *Criticism and the growth of knowledge,* Imre Lakatos and Alan Musgrave, (edit), Cambridge University Press p 265)

[135] Kuhn, T. (1970) Reflexions on my Critics, in: *Criticism and the growth of knowledge,* Imre Lakatos and Alan Musgrave, (edit), Cambridge University Press, p 265

make, only thanks to his identification of shared with true knowledge, he can conclude from the humble progress towards corrected knowledge, that any knowledge in the natural sciences must be also ever only relative knowledge, just what the dogma of approximating knowledge in the social science wants to say, phrased—in Kuhn's variation—as the way how knowledge progresses, in Popper's way as how thinking processes.

Whatever Kuhn tried to clarify, dethroning the objective knowledge of the natural sciences towards also relative knowledge is what his theory about the progress of natural science knowledge is saying, may this be with our without the notion of a paradigm shift and dethroning the envied hero of natural science knowledge was an essential element needed for the post-structuralist attack on the remaining ruins of scientificy in the social sciences, thus paving the ways for the funeral of science and for advocating precisely the relativism of theorizing, justifying all the patriotic theories as coinciding with the latest epistemological insights of the social science gurus.

Now we finally know that "The structure of scientific revolutions" is the revolution to subsume natural sciences under the epistemological dogma of the social sciences and why this theory about the progress of knowledge in the natural sciences could become a real paradigm among social scientist, who did not only need to understand any natural sciences, but who appropriated this knowledge as their paradigm about any thinking in any science. Without the delays, natural science knowledge progress needs, the promotors of the social science dogma, insisting that there is no objective knowledge, did not need any circumstances, any paradigm shift, allowing them to arrive at sharing this paradigm about the relativisms of knowledge, since these insights served form a natural scientists about the natural sciences is what they already knew as the only secure knowledge social sciences know and allow, much before the notion of a paradigm shift was found.

In other words, the fact that this insight about the natural sciences, which ever so painfully questioned with their knowledge the dogma of social sciences that there is no objective knowledge, thankfully presented by a natural scientist theorizing about the natural sciences, made a paradigm shift without shifting a paradigm, just because it was this false conclusion T. Kuhn provided, that was confirming, what social sciences ever knew was a correct theory about theorizing, now liberated from only counting for the social sciences.

This, proving the impossibility of knowledge with the progress of knowledge, was a masterful piece of social science thinking, that did not need any major circumstances to be acknowledged by the social sciences and by the natural sciences thinking about theorizing, and this without any revolutionary delays, since it did not say anything revolutionary for social sciences, but provided a theory that compliments the epistemological relativism of all the Poppers with a relativism attached to the envied and hated memorial the social sciences saw in the knowledge of the natural sciences.

It was precisely this debate that abolished the natural sciences as the final barrier for the proclamation of a concept of all sciences, in which the subjectivity of knowledge was from there on discussed as liberating science from what the social sciences had coined the "Newtonian" rationalism, a way of theorizing that insisted on the objective of scientific thinking, creating knowledge that claims knowing the world unlike subjective opinions, may this be even in its most relativated version of Poppers idea of an ever approximation towards knowledge. Even in this skepticist version though also opposing objective knowledge, the never reachable ideal of objective knowledge is at least kept as this very ideal that scientific knowledge is striving for objective knowledge, but can only ever approximate it. It was the idea, opposing this mere ideal, for which the debate about T. Kuhn's historical knowledge relativism was instrumentalized, to finally liberate science from what science distinguishes from any subjective opinion, thus providing the very epistemological justification for the concept of the mysticism of spatiological and patriotic thinking, allowing with the epistemological gurus of post structuralists theories to create the absurdity of contextualized theories for the sake of creating knowledge that represents spatiological knowledge, re-presenting any local, patriotic view as a spatial contribution to the global knowledge arena of spatiologically constructed theories.

It could not bother the social sciences, that replacing scientific thinking by spatiological theories, the relativisms of scientific knowledge with the new dogma of contextualized thinking also abolished scientific knowledge as a voice in any global debates about the world's social. How could any scientific view be counted as a distinguished view, distinguished from views on any phenomenon phrased by political or economically biased voices, if sciences admittedly considered scientific knowledge as being also nothing else but any subjective knowledge, just as the explicit interested views of the political and economic elite as the same kind of knowledge as the scientific knowledge.

Thus, for the price of sacrificing science as a distinctive voice in the global discourses about the world's social, scientific knowledge in its version of a contexualized thinking was set free to articulate any spatiological knowledge as—from there on—a scientific contribution to global social thought under the regime of the social sciences. It was this instrumentalisation of the discourse about T. Kuhn's historical relativation of knowledge that provided the epistemological justification for all the battles between spatiologically constructed knowledge, global versus local, northern versus southern and whatever the local entities in these battles among spatiological knowledge may be. It was this debate that invited arguing against scientificy as the last hurdle for practicing global social thought under the social sciences as a discourse among the many spatiological knowledges, for social sciences no longer searching for knowledge but searching for the moral values of an "universal universalism" that must be found by science beyond science.

Scientific knowledge thus was made one moral opinion among others, a contribution devalued from the power of objective knowledge, another case where the campaign against inequalities succeeded and made scientific knowledge across the world equal knowledge, including the knowledge articulated by the political and the economic elites—with the slight difference compared to the scientists, that the political and economic elite knowledge holders, are not only holding one of the many relative knowledges, but own the power to set in force what they think, while they let the others share all their irrelevant views anyway questioning each other via their relativism in their global discourses about all sorts of patriotic knowledges.

And this is very instructive, firstly, what this global discourse among the many spatiologically relativated knowledges is about and, secondly, how such a pluralism of subjective, spatialized knowledges, not in the natural, but in the social sciences, sciences which, unlike the natural sciences, do not want to progress from false to right knowledge, how these social science knowledges progress—progress from where to where?

Chapter D
The discourse about and the progress of social science knowledge

If there is one thing that one cannot critique the social sciences for, then it is that they are not critical. Discourses, critiquing theories in social sciences is not only activity accompanying the creation of knowledge, discourses—according to social science epistemological insights about social sciences—discourses about knowledge even are the means to distinguish which knowledge is true knowledge and which not—just as T. Kuhn made the progress from false to right knowledge, his paradigm shift, a matter of "*what the members of a scientific community, and they alone, share.*" And this raises at least two questions: How do social sciences, who do not want to distinguish between false and right knowledge, share knowledge and how does social science knowledge progress?

"Critical thinking" is considered as the core ability of academics for the distinction between—between what?

> "Critical reading, writing, speaking, and listening are academically essential modes of learning."[136]

Nonetheless, how does this critique work if social thought under the social science theorizing must be constructed from assumptions and, thus, can only ever arrive at knowledge constructed from any assumptions, the ex-ante chosen theoretical frameworks through which they theorize, what does a science critically argue about, if it cannot argue about the question if any knowledge is wrong or right? More precisely, if wrong or right knowledge are relative criteria for knowledge, relative in the sense that if knowledge is wrong or right depends on the theoretical framework through which this thinking theorizes, what do discourses about relative knowledge discuss?

And, secondly, if critique is a must, an essential of scientific thinking, a dogma that ever applies to any knowledge, what is a cri-

---

[136] http://www.criticalthinking.org/pages/the-national-council-for-excellence-in-critical-thinking/406

tique about that is not only a must for the debate about any theory, but an "essential mode" of social science theorizing?

And, thirdly, what is the nature of the social science discourse about knowledge, the way critique is made, if any knowledge constructed through theoretical frameworks must be both accepted as knowledge and always critiqued?

Finally, what is this discourse aiming at, if it ever must criticize knowledge and cannot arrive at excluding false from right knowledge and to thus progress with the creation of knowledge that eliminates false from right theories, just as if is the case in the natural sciences? How does social science knowledge then progress?

The discursive creation of acknowledged true knowledge

If one asks Habermas, certainly one of *the* global experts in reflecting on discourse, who does not only think that theorizing is all about communication, but even believes that the whole society is a matter of communication, one gets the following answer

> "If there are any doubts, the difference between true and false knowledge must be decided via a discourse."[137]

One can do many things in disputes about theories and one can certainly also distinguish between false and right knowledge and it is this why it is remarkable that social science thinkers, who advocate that knowledge must ever be only relatively true knowledge, while arguing about the discourse about knowledge they know very well that there is obviously a *"difference between true and false knowledge"* and that they also know very well that it is this difference that decides essentially on what knowledge is and what is not. But then, if discourses are able to make this distinction, why not right away distinguish between false and right knowledge along the knowledge lines, especially *"if there are any doubts"*? Why only distinguishing between both only ex-post *"via discourse"*?

---

[137] Habermas, J., (1974) Preparatory remarks about a theory about communicative competence, in: Habermas, J.; Luhmann, N., *Theory of Society or Social Technology?*, Frankfurt a.M., p 135 (own translation)

And, more importantly, how can knowledge that can never make this distinction along knowledge between wrong and right, since any knowledge is always wrong and right depending on the theoretical framework through which it is constructed, how can social sciences which cannot distinguish between wrong or right and eliminate false from right knowledge, how can they exactly do this via a discourse, if none of the discourse participants can know what false and true knowledge is? This is mysterious.

And it is even more mysterious, how a science, which strictly rejects to distinguish between true or false knowledge, should be able to arrive at distinguishing false from right knowledge via a decision among scientist? How do they decide what is wrong and right if wrong and right knowledge cannot be created by thinking, but now is supposed to be decided "via a discourse" among all these social science thinkers, who can only create relatively right or wrong theories?

Habermas knows the answer:

> "The idea of a true consensus provides from the participants of a discourse to have the competencies to judge about the truth of statements and the correctness of activities."[138]

"The idea of a consensus" is however no way out of any false knowledge: A consensus can also easily be made about wrong knowledge. As one can see not only from the Copernicus example, where theorists shared a consensus on false theories, the fact that all discussants agree about knowledge is no proof of knowledge, but only an agreement about what the discussants *decided* as being knowledge—and the history of science is the history of false knowledge, the more if it is a matter of decisions on what is true and what is false knowledge.

And the competencies of scientists are even more dubious to make such decisions, if these are scientists who all can, due to the epistemological dogma of social sciences theorizing, only create knowledge that can only be relatively right or relatively wrong, but in a discourse are now supposed to have the competence to decide what is right and wrong knowledge? The question remains, what the competence social scientists have is, who cannot distinguish between false and true knowledge "to judge about the truth" while

---

[138] Ibid (own translation)

judging about knowledge, but who are able to distinguish what the true knowledge is "via discourses", a distinction they make via competent decisions? What is this competence social scientists have the social sciences cannot have arriving at a *"consensus" "about the truth of statements and the correctness of activities?* What is a consensus about *"the truth of statements"* in social sciences which can only create relatively true knowledge?

An example of how social scientists argue about theories, how they discuss theories, will disclose the secret of what a true theory in social sciences is that does not know any true knowledge but that can decide via a discourse what true knowledge is.

> "We have used the instruments of science; we have counted and measured and compared; but something essential to scientific practice is missing in almost all current discussions about human behavior. It has to do with our treatment of the causes of human behavior....Yet almost everyone attributes human behavior to intentions, purposes, aims and goals....What is needed is a technology of behavior"[139]

On not less than 25 pages arguing about the need of a "technology of behavior" the founder of behaviorism tell us that the whole history of sciences did not explain what this scientist feels needs to be explained via a theory of behaviorism. Doing this, not a single false thought is critiqued, not a single false explanation in all the sciences and theories that are mentioned, are disproved, but the one "proof" that no social science theory ever created Skinner's theory, a theory explaining the behavior of human mind, yes, with a theory interpreting human mind as—human behavior.

Skinners way of arguing is typical for social sciences which cannot argue that any thought is wrong, or made any mistakes that the critic will now eliminate, since they cannot create theories that are right. The "technology" or arguing in social science discourse is therefore negative, negative in the sense that it does not disprove any thought, but it discredits thought as not sharing—the critic's thought. The mistake of the existing knowledge is that it is not the knowledge, the critic adds to the existing knowledge he accuses of to not being the knowledge he has.

This is how social sciences discourses argue and how they can only argue: They *denounce* other theories of either reflecting

---

[139] B. F. Skinner, (2012) *Beyond Freedom...* p 7–22

through the wrong theory or with the wrong methods for the theory they choose, that they forgot the aspect another scientist did not forget and so on and on. Allowing each other to boast with their expertise about the endless combinations the methodological apparatus and its rules offers a debate that surely never does one thing, that it touches the coherence of the contents of any thoughts. Juggling with theories, made towards approaches, and with methods made for theorizing is the art of discourse in the social sciences, denouncing other theories as not being the one's the critic favorable theory, approach or method, or—better—ones favorable combination of all, boasting about an own "complex" combination of methods or new approaches, combining theories towards presuppositions nobody has made before, thus never arguing against any thought and, instead, the repetition and referencing of knowledge one has merged to a new never before existing combination of thought, is the art of arguing in the social science discourse, that does not trace any thought in any theory. Arguing this way, it is guaranteed that any construction of thought, their observations, conclusions, in short, any of their cognitive operations is never discussed, but ever discredited as not doing, what the criticist does.

Social science discourses ever critiquing and never questioning the argumentative structure of thought thus immunize social science thought against any critique and it is in fact this concept of critiques that is the strongest weapon of social sciences against any critique of their theories.

Stating that what has not yet been said with no other critique of any existing theories but by accusing them of not having created the critic's theory, hence, the glorification of one owns theory, is the other side of the coin of discrediting other theories, and it this, the nature of social sciences discourse through which they arrive at the main objective of discourses, glorifying ones theory as the one and only theory and doing this is the competence Habermas advocates, enabling social scientists to distinguish between false and true knowledge via a competent decision in discourses that introduces a distinction of false and true knowledge as a *hierarchy of theories,* on which social science theorizing is based, theories that are only theories and other theories which are the "grand" theories for theorizing, providing the "approaches" through which theorizing constructs theories.

Who could give a better example for what this discursive competence is, but the pope of discourse theories. Habermas himself provides not only an instructive case of what the discourse among all the relativated theories is finally all about and what the secret of those "competences" is that solves the problems of a discourse among discussants who all can not own any right knowledge but have the competence to distinguish between false and right knowledge, in other words, Habermas reveals what truly true knowledge is, found as a consensus "via discourses" among the competent social scientist:

> "From a research strategy perspective the value theory (of Marx, M.K.) aims at solving the problem making a systemic integration visible on the level of social integration:"[140]

Habermas calls his discussion about Marx's Theory *"Understanding Marx better, than he could understand himself"*[141]. Habermas's competent discourse with Habermas about Habermas's interpretation of Marx' theory, a theory about capitalism, the classes and their conflicting interests, results in this statement: Habermas interprets Marx' theory about antagonistic classes as an attempt to say—of course, what else—what Habermas's theory about social integration says. We now understand Marx theory better, once we understand his theory as a try to say what the competent discourse expert Habermas says: Understanding Marx better by interpreting him as a predecessor of Habermas. And not only Marx, but basically the genealogy of social sciences, or better those to whom Habermas compliments the honor to appear as predecessors of Habermas. And this is, indeed, the whole truth about what the competence and what the discourse is, via which social sciences clarify *"the difference between true and false knowledge."*

Though not many scientists consider their theories as the final climax of social science theorizing, Habermas's self-enthronement as a mega thinkers leads us to the answer what and how social science decide by their discourse of the non-knowers what is true knowledge.

---

[140] J. Habermas, J. (1976) *Zur Rekonstruktion des Historischen Materialismus*, Frankfurt a.M., Suhrkamp, p 223 (Own translation)
[141] J. Habermas.,(1963) *Theorie und Praxis, Sozialphilosophische Studien*, Frankfurt a.M., p 244 (Own translation)

Making not only Marx but the whole history of social sciences predecessors of Habermas' theories and their theories only understandable through Habermas, just as skinner introduces his theory about behaviorism as a new episode of theorizing, are not at all reflexive accidents of the particular arrogance of these true knowledge creators, the meta—theorist's theories through which the others theorize. Critique in the social sciences is about arguing about the non-arguable, since it is the choice of a thinker to choose any theory he prefers to choose for theorizing. The true substance of the discourse in the social sciences therefore is to decide via the discourse among the social scientists which theory they attribute the status of a model theory for theorizing and, hence, to which scientist they attribute the position of a scientific judge to decide about this. Downgrading other theories—in our case—Marx's theories towards a mere theory that needs Habermas's theories to be understood, is the way to create *hierarchies of social thought*, in which the "grand" theories enable theorizing to distinguish between false and right knowledge as knowledge that is *ruling* theorizing and the other knowledge as a mere application of the "grand" social thought. Not having the slightest clue about what capitalism is, but enthroning his own thought as a theory through which theories can only be understood is the cognitive operation how "grand" theories are created as deciding about how to distinguish between false and right theories as theories, which get the blessings of the self-enthroned theories of being constructed through those master theories. What often looks as if it was a sort of juggling with theories is, as in our case Habermas, the attempt to enthrone the theory juggler as such a meta-thinker and his theories as theories which rule the cognition in social science thinking in which thinking is thinking through such ex-ante theories—which also shows the dilemma that to finally become such a ruling meta-theory it needs other thinkers of the same kind who construct their true theories with the proof of their thought by thinking though ruling theories.

Social science thinking necessarily externalizes "truth" from the substance of thoughts towards the subjects of thoughts, and invites the social status of the subjects, their "competencies", proved by their means of "scientific power" as a criterion for which knowledge counts. Social science thinking materializes in its discourse modes the *democratic* nature of the social sciences epistemology, finally simply *defining* that true knowledge is knowledge that gains its status of true knowledge through the acceptance in such critical discourses, the acknowledgement knowledge gains by the social sciences determining what counts as knowledge, and thus, materialize the paradox of an approach to social science thinking that practices thinking as voicing the non-comprehended nature of their objects of thought as mere conviction among social scientists. A theory that is quoted as knowledge must be knowledge and a theory that is quoted as knowledge must be quoted, is the secret of finding truth in the discourse among the competent social science thinkers. Which social thoughts gain the status of a grand theory are those social thought shared by the scientists as "grand theories", entitled to rule thinking as a theoretical framework through which the presupposed way of social science theorizing constructs theories.

Science indices, scientific excellence and the like are the categories materializing this very objective of the social science discourse about the many presupposed theories, the telling categorical instruments to distinguish false from true knowledge via Habermas's discourse among the competent discussants as the distinction of the many theories, between theorizing as applying those theories acknowledged by the discourses of the competent discussants and the grand theories framing theorizing.

Establishing a theory as a reflexive instance that rules thinking and through which other theories can only be understood, is what the social sciences establish through their "competent" discourse in an approach to science in which knowledge consists of the modelled theory through which it is constructed.[142]

One should therefore not misunderstand that Habermas's unbeatable arrogance is, indeed, a personal arrogance; this arrogance is, however, the inevitable consequence of a discourse that cannot critique any knowledge other than opposing it for the theoretical framework it applied or the methods it used, never the knowledge

---

[142] Science indices materialise this as a decision technique. What counts as "excellent" knowledge and what as inferior knowledge is decided by the academia via the number of quotations, when they distinguish via their discourses—and not only them, as Habermas insists—"*between true and false knowledge*". What could better document how irrelevant the content of knowledge in social science discourses is, when they distinguish between false and right knowledge!

it discusses; and to enthrone one's own theory as a theory, "coining" a scientific currency, that rules thinking is the inevitable arrogance of a mode of critique within the approach of theorizing about modelled thought that can, consequently, only argue about which theory achieves the status of a model for thinking in a hierarchies of knowledge. How else than with arrogance of mere *decisions* and the according submissiveness of those who indeed "share" these decisions could a theory among all the modelled thoughts claim to be the judge for and over all other theories?[143]

This, the non-discourse of the social sciences, never critiquing, but denouncing other theories for not using this but not another theory as "framing" theorizing, it is this arrogance that gives birth to what is falsely described and discussed under the notion of "scientific power". It is falsely described, since it considers denouncement as a matter of personal style of "Western" scholars, or, more prominently, as a matter of any power means to impose a theory as ruling theorizing.

However, this difference between making both arrogance and ignorance a necessity of the social science discourse and scientific power is important. It is important, because it is this difference, that prevents the opponents of what they oppose as scientific power from critiquing the theories they oppose and, instead of critiquing them they rather discuss some "unequal" conditions for never gaining the status of a ruling theory. The notion of scientific power accuses the claim of other theories to model theorizing and does not want to notice that this requires the acknowledgement also of those to accept such a theory as a "coining" theory, an acknowledgment these criticists contribute —ironically— via their critique that their theories never get this degree of acknowledgement. Gaining this status needs the acknowledgement of scientists uncritically applying such a theory to theorizing and only thus, by using theories as model theories, they establish this theory as a theory that

---

[143] It should be mentioned that it is exactly the same Habermas, who insists that the discourse that decides with the competence of all the Habermasses, which theory gains the status of a ruling theory, can only be carried out by academics whose competencies are legitimized as Western democrats: *"To principally distinguish a dogmatic acknowledgement from a true consensus, a universal and non-ruled communication is pre-conditional."* Habermas, J. (1971) Der Universalitätsanspruch der Hermeneutik, in: *Hermeneutik und Ideologiekritik,* Frankfurt, p 156 (Own translation)

rules social science theorizing and the criticists critiquing the unequal distribution of the scientific power thus add their acknowledgment to the theories they do not want to oppose as theories via this very type of critique.

Establishing a theory as a meta-theory is the objective of this type of discourse on either part, not the action of any individual social scientist. This, creating a *hierarchy of social thought,* is the result of the social sciences discourse and of their style of debating—and discloses what a scientific discourse is about, when such social science discourse discusses and critiques the many nation state theories to arrive at *globally* acknowledged knowledge that *rules the creation of social thought across the world.*

Paradoxes of acknowledged knowledge in the global social science discourse

Globally accommodated to social science thinking, the social sciences across the world theorize about state islands socials, reflected on through the very idealized nation state views and thereby contribute theories to the global discourse among spatiologically constructed knowledges, emphasizing the clandestine, authentic nature of their patriotic theories, emphasizing thinking as thinking bound to the pre-assumptions of the "contexts", the idealized state views on the state socials, created through the meta—theories of disciplinary thinking and the admitted anti-scientificy of all the "authentic" knowledges, all these theories are populating the global social science world and create the paradoxes of its discourses.

> "These difficulties are not only due to the difference between English and French. They probably also reflect the French conception of knowledge, which puts an emphasis on explicit and scientific knowledge, and the French conception of learning, which traditionally puts the emphasis on formal education and training."[144]

Unlike discourses within the same nation state science about which knowledge rules theorizing, the euphemism of the so called "national science communities „phrases so neatly as if it has anything

---

[144] Mehault, P., (2007) Knowledge Economy, Learning Society and Lifelong Learning—A Review of the French Literature, in: Kuhn, M., *New Society Models for a New Millenium,* Lang New York, p 67

else in-common but this battle about acknowledged knowledge, starting from categories constructed from the same national social and from shared meta-theories, discourses among different nation state science theories manoeuver their spatiological knowledge into an impossibility to even pretend to share knowledge as reflecting on what they say.

Trapped by the illusion that the nationally particular way nation states shape their policy domains all nation states have in-common, here education, was any substantially peculiar way of education only this nation knows, the "difficulties" to share his spatial knowledge about education in France this scholar discusses, is the difficulty social science thinking that is conceptually thinking in nation state rationales—inevitable—encounters, once it enter discourses beyond the national science borders and debates with other sciences of the same kind of theories created though such national theoretical constructs

The fact that education in France teaches what Mehault calls "scientific knowledge" appears to such nation state science thinkers as if there was a "*French conception of knowledge*". Detecting the peculiarities of the "French conception of knowledge" as a difficulty for other biotopic thinkers of the same kind from other national socials to understand education in France can only be detected by thinking that cannot think about the social other but through the idealized nation state rationales as disciplinary thinking does, that is therefore so much caught in the national peculiar interpretation of a sphere of state policies, that this way of scientific thinking seriously creates the notion of a national concept of knowledge and that one must share this national construct as a precondition to share theories about this nation state social.

For this kind of state science thinking sharing the peculiar interpretations of a nation state rationale is the cognitive condition to theorize about the national biotopes. Being introduced into the secrets of the national peculiarities of the inhabitants of the biotopes state science thinking considers as the cognitive key to reflect on each individual state social enclave—a cognitive circle that results in a monstrous circular global discourse, the incomprehensible tautology of a diversity of patriotic knowledges that can only be understood under the precondition that one shares the national categories only thinkers can understand who share the national context that creates them.

Global discourse about acknowledged knowledge ruling social science theorizing

However, global discourses among the spatiologically constructed social science theories do not only have any major problem, making it a precondition for sharing their knowledge to share their exclusive national views about their nation state social. Obviously, one must conclude that sharing knowledge among the national knowledge islands is not really the objective of the global social science discourse.

The universalization of the theories created through social science thinking in the imperial world has, in fact, received a response that not only very much shares their approach that social thought must be thinking on and through a nation state view perspective, that theorizing must be bound to thinking through and for nation state rationales, but also that the global discourse must be about nothing but the question which spatiological theories rule the global creation of knowledge and its discourse.

> "The study of Latin America's economic life is primarily the concern of its own economists. Only if this regional economy can be explained rationally and with scientific objectivity, can effective proposals for practical action be achieved."[145]

Not only is creating spatiological knowledge a contribution to a discourse which is not about the knowledge others create, global social science thinking about global social sciences discourses goes one step further: For the sake of a global discourse nationally "biased" theories are made the precondition to join the discourse and reveal what the substance of a global discourse among such theories is about:

> "Sarkar looked at the history of Asiatic sociology and compared Sino-Japanese Buddhism and modern Hinduism. He argued that Buddhism in China and Japan had its origin in Tantric and Pauranic Hinduism.
>
> The Hindu or nationalist bias is hard to avoid in this example, but more important for our purpose is the attempt at developing non-Western theories to study local realities."[146]

---

[145] Prebisch, Raul. (1999) *The economic development of Latin America and its Principle Problems.* United Nations, Department of Economic Affairs. Reprinted in David Greenway and C.W. Morgan (eds.), The Economics of Commodity Markets, Cheltenham, Elgar, p 7

What a hypocrisy: In fact, if a theory aims at developing theories, at social thought constructed through the views of "local realities", bending social thought of the kind of Weber's "*Wirklichkeitswissenschaft*" under the rationales of individual state "realities", then "it is hard to avoid"—that they do what they are aiming at. And then, it is "hard to avoid" that the encounter of nation state science theories, theories constructed about and through the rationales of individual state socials, encounter each other as non-comprehensible global social thought considering the very non-comprehensibility of theories as the way to join this discourse.

State science theorizing, that creates purposefully "biased" theories for the sake of a discourse among the global knowledge islands, obviously aims at anything else but sharing the confined knowledge of state science thinking, aims at anything else but understanding what is going on in the other nation state socials.

> "Sociology is most commonly classified in the West into two trends, depending on the starting point: the individual or society, the partial or the whole. This classification is confusing to many Arab researchers who have not found mature European individualism in their societies or coherent nations made up of Western-like stratified societies from which they can start their research. Maybe that is why Arab sociologists believe "Western" sociology is not suitable for their societies and resort to the Arab-Islamic heritage to come up with a "local" sociology. (Ali al-Wardi's work is an example that we will later discuss in detail). A new Arab proposal in this regard is gaining prominence; it calls for a new classification of sociology into two trends: balance and conflict. A leader of such a classification is El-Sayed Yassin, and the classification itself is an attempt to rescue sociology from attacks by critics who see no use in it."[147]

The least that seemingly comes to the mind of this "Third World" scholar is to critique *the* false dichotomy on which sociological thinking is based and to oppose social thought created by an affirmative states science thinking in the Western knowledge islands. That theories do not explain what they are about, but a model through which the social reality is interpreted, the nature of thinking in models, this way of social science thinking is so natural to

---

[146] S. F. Alatas, (2010) The call for alternative discourses in Asian social sciences, in. UNESCO Publishing, *World Social Science Report, Knowledge divides*, Paris, p 172

[147] Faleh Abdel-Jabbar, (2013) *Insights into the Topic: Arabs and Sociology, Epistemological and Ideological, Characteristic and Seclusion, Synthesis and Openness*, 2013, unpublished paper, p 1

thinkers opposing the "Western" theories, that they feel that these theories are "not suitable" and seek for another model of preoccupied thinking. Rather than critiquing any of the opposed Western theories, they find thought that duplicates a social into a dichotomous opposition of the individual members of a society with themselves as the whole society very logical and just feel that the less mature societies in the Arab world would be better off with a variation of this false very theory accommodated to an Arab state social thinking. In fact, the idea to replace "society versus individuals" by "balance versus conflict" is genial, because it articulates precisely the same, since balance and conflict are the procedural interpretation of what the relations of what "society versus subject" are all about: According to the ideals of sociological thinking it is the society that must keep the *balance* between the *conflicting* interests of their individual society members. Nevertheless, "*the classification itself is an attempt to rescue sociology from attacks by critics who see no use in it.*" It is, indeed. But why is it so important to rescue a theory, if "*Arab sociologists believe "Western" sociology is not suitable for their societies*"?

A Japanese sociological thinker also discussing the non-suitable "Western" theories helps to understand what the whole discourse among purposefully "biased" state science theories and the global discourses among them is all about.

> "Why is social capital, rather than aidagara and en, popular among sociologists worldwide, even though the terms are similar?"[148]

> "The second strategy is a particularism-to-universalism-to-particularism strategy. Using this strategy, the Japanese sociologist would invent a broad concept that covers both Japanese and the Western types of social relations, as well as the Chinese type. The Japanese sociologists could then derive local concepts such as aidagara and gunaxi from the more general concept. Generally, this is an authentic scientific strategy and is, therefore, preferable."[149]

The global social science discourse among the confined knowledge islands is not at all about compensating the knowledge deficits of knowledge confined to the local knowledge islands, but about the question of how the nationally constructed theories manage to set

---

[148] Y. Sato, Are Asian Sociologies Possible? Universalism versus Particularism, in M. Burawoy, M. Chang, and M.F. Hsieh (eds), *Facing an Unequal World: Challenges for a Global Sociology*, Vol. 2, Asia Taiwan, p 193

[149] Ibid, p 199

their own reflexive models for theorizing about the secluded state socials as the reflexive global standards, the global model for the pre-assumptive thinking through models. Discourses among the "authentic" state theories do not care about knowledge, neither in their own knowledge islands nor in those abroad. The *"particularism-to-universalism-to-particularism strategy"* is about creating a "local" theory, universalizing this theory as a global model for theorizing and then using this universalized local theory as the world's reigning theory for theorizing—just as the Western prototype of colonized state science thinking has done.

To achieve this aim, setting a national social science theory as a model for global social science thinking, social science thinking, arguing so passionately against scientific power, inequality and the like does not hesitate to sympathize with the real imperial power as a vehicle to imperialize national social thought:

> "Here it would be interesting to speculate about how academic dependency may be affected by the shifts in the balance of economic power. It is not uncommon in Asia to hear optimistic views to the effect that if Asian economies overtake the West, Asian culture will become more dominant globally:...But, it is doubtful that any Asian nation or Asia as a whole would become dominant in the social sciences on a global scale. The case of Japan is instructive in this case. Japan is a world economy power but it is not a social science power by any means."[150]

Nothing but this critique of the non-suitable Western theories and the search for alternative theory models for theorizing are further away from the question, if the opposed theories creates any knowledge about the world. They just pretend being a critique of the theories they oppose and the discourse among social science theories are only concerned with a scientific battle in which they seek to question the ruling status Western theories have gained as models for pre-assumptive global social science thinking and to complement their idealizing way of nation state science theorizing with a multiplicity of nationally "biased" ways of nation state science theorizing, if possible making them with the support from a most realistic imperial political power a global ruling theory, ruling the global creation of social science theories.

---

[150] S.F. Alatas, *Academic Dependency and the Global Division of Labour in the Social Sciences,* Current Sociology 51, p 605

All the controversies about knowledge flows, academic dependence, inequalities, peripheries and centres and the like are the accompanying debates of a scientific enterprise that has replaced the question of how to explain the world by the question how to "coin" social thought ruling pre-assumptive theorizing. Only a science that argues about the definition power for thinking, not at all about theories, needs to argue about all the power means, the where, the who is who, flagships, rankings, in-equalities, "scientific imperialism" and the like. A science that was interested in knowledge would just need to discuss one question, that is if any theory where ever it is created is correct or not, no matter who, where, when and about what it is created.

Critically reflecting on theories and raising the question about who dominates the global discourses are opposing issues.

> "I would argue that thin concepts spread faster among sociologists than thick concepts, which are loaded with local meanings. This is because when individuals are exposed to a concept and try to understand it, a thin concept has lighter cognitive burdens on its receivers than a thick concept"[151]

Social science thinking especially in the—therefore rightly called—*inter—national* discourses is interested in anything but knowledge, and one could get the impression that arguing about the definition power, defining the models through which the social must be reflected across the world, "the thin and fast spreading concepts", liberates state science thinking from all the superfluous "thick" scientific attitudes of scientific thinking and allows them to arrive at the final substance of the pre-assumptive social science thinking in its discourse, in a dispute about the question which of the idealized versions of nation state theories rules global social science theorizing about nation state socials.

Arguing about the position national knowledge bodies hold

While the discourse about which knowledge from which national way of theorizing is acknowledged knowledge and thus rules thinking across the world still *argues about the theoretical substance* of

---

[151] Y. Sato, *Are Asian*....p 197

theories, the discourse about which knowledge is acknowledged to rule thinking across the world, is accompanied by a further development of the discourse towards a debate, in which the theoretical substance of theories no longer matters and is replaced by a debate about the sheer means to "dominate" thinking beyond any reflections on what theories say.

The least this social science discourse seems to be interested in is a discourse about and among the nationally constructed bodies of knowledge. The number of citations, rankings, excellence, credits and alike are indicating what social thought in this social sciences discourse is concerned with. Knowledge is not a matter of what it tells us about the world's social, but a matter of representing a position national knowledge bodies hold among a multiplicity of spatially distinguished knowledges. In this discourse they argue, not about controversies between knowledge originating from different parts of the world, but about the status between such entities such as "Southern" and "Northern" knowledge, knowledge from scientific centres and peripheries and the like, *knowledge representing* all kind of *national*, local and global *knowledge bodies*, in short knowledge that distinguishes social thought as *representing politically defined spaces,* discussing about the position the spatially distinguished knowledge bodies hold and the means they own in a global battle among these national knowledge bodies, their "scientific power", "dominating" knowledge, identified and measured via criteria, signaling the means for a kind of reflexive command spatially distinguished knowledge has over the world's assembled knowledge bodies, engaged in arguing about knowledge "hegemonies" among local, glocal and global knowledges and their means to join these battles about "scientific power". Politically defined and spatially distinguished knowledge bodies, identified via the nationality of the thinkers subordinated, whatever they say, into to these bodies thanks to their nationality, all populating a global knowledge arena, arguing about the scientific positions the knowledge bodies they represents gain or lose in a global knowledge battle. The losers of this race discuss a science world, competing about the status these knowledge bodies holds in a global science arena as a battlefield among "foreign", "dominating" or "imperial" thought, thus sharing the view that global social thought is finally not about knowledge, explaining the world's social, but that knowledge is finally nothing else but a weapon to decide among a multiplicity of competing politically defined knowledge bodies about the power they have to decide about the status theories about the world's social hold for theorizing.

And as much as social sciences seemingly are concerned about knowledge as a means to "dominate" other knowledge bodies in the world, the prevailing critique of theories this critical image about the social science world has, seemingly, is not interested in understanding the reasons for such an odd science world arguing about such things like a *"scientific imperialism"*, but rather echoes the outcomes of this discourse about ruling the world's national knowledge bodies, mainly phrased from the perspectives of the losers, and argues against not owning the means to join this battle, heavily question the criteria for their "rankings", the "in-equalities" or "asymmetries"—a more recent category, created by US militaries in their 'wars against terror' in Iraq—of their publication policies and the like and all sorts of the lacking conditions for participating in the global social science arena and the race among about subjects they for this very purpose make up as "national science communities, opposed as a "scientific imperialism", passionately questioning the unfair rules and means of these battles, these opponents are apparently keen to be part of, not realizing how much they *are* part of it with this very opposition.

Rather than critiquing social thought that enters the global social science arena as nationally constructed knowledge and argues about the status of nationally constructed knowledge bodies, the mainstream opposition argues that this discourse does not give all those nationally constructed theories the same means to participate in the global knowledge battles about "scientific power".

Arguing about the position their knowledge holds in the global knowledge arena, the global social knowledge arena and their disputes are populated by subjects like "flagships" and "national sciences communities", all in a battle about their position and means to "dominate" theorizing, a battle about "scientific power" which is critically observed by some scientific arbitraries, who categorize the players of this global battle in centres and peripheries, Western, Northern and Southern theories, or distinguish them in "scientific powers", and "dependent" science communities.

Instead critiquing the nationally pre-occupied views of a nationally confined social knowledge and their discourses about which national knowledge body rules thinking about the world's social, these opposing views are concerned about the distribution of the scientific weapons in the battle among ever nationally constructed entities about the "scientific power" to rule and to intervene into the status they have among such national knowledge bodies.

In this critical view of opposing global social thought as a controversy among nationally constructed knowledge subjects about their status as a ruling or ruled theory, a controversy in which theoreticians about the world's social appear as experts with knowledge about national socials and are primarily considered as representatives of nationally or global constructed knowledge entities, whatever their theories are about and whatever they say, it is this opposition that does not oppose the constructs of social sciences theorizing, claiming the world's social could be understand by an agglomeration of thinking about secluded national socials. Instead opposing these false assumptions of social science theorizing and of a discourse, that is keen on arguing about the status of their invented national knowledge bodies—the sociological invention of "national science communities"—these debates about who has the say in global social sciences proceed to distinguish social thought no longer according to the simple national belongingness of thinkers, but further develop this towards creating global social science entities, such as a "global North" and a "Global South", imagined in a battle among these global politically constructed entities, gathering theories in those regionalized knowledge bodies, in which the view of the gathered theories are simply ignored and assembled via an assemblage of those national science communities towards these global knowledge entities and their imagined battles about a the status of theories "dominating" global theorizing, in which the least that matters is what these theories say.

Within these global encounters of nationally/spatially constructed global knowledge bodies, knowledge is so naturally seen as a complementary means for the battles among these political global knowledge bodies that the contents of thoughts have become rather an accessory for identifying the nationality or the global belongingness to world regions of social thought. Practicing a kind of global scientific racism judging about theories due to the nationally defined location where it is created, is considered as an argument against theories that only needs to argue about theories by identifying the space, i.e. national "context", in which it was produced, to thus know, which theory is the rightly critiqued and who is the right critic.

The "Northern" social sciences are opposed by an alternative plea for a global "provincialisation" of theorizing, never critiquing, but advocating to downplay the theories from those knowledge islands in the "Global North" towards also only mere local knowledge, not rejecting the idea of knowledge that rules thinking, but rejecting their claim to monopolize ruling the world's social thought, downplaying, not critiquing, this knowledge to only one provincial theory among the many locally confined knowledges in a science world that, for these criticists obviously also rightly consists of an agglomeration of the many provincialized social thought and their battles about the status of their patriotic theories dominating thinking or being dominate. Beyond this dreary dichotomy to either dominate or being dominated there is seemingly no alternative. Sharing the social science concept of nationally constructed social thought as a contribution to represent global knowledge bodies this counter-discretization of theories rejects, not critiques the claim to rule global social theorizing and thus reveals that these incriminated "Western sciences" are not only practiced in a particular provincial space of knowledge production *in* the Western part of this globe, but have become a mode of thinking and debating, universalized and practiced across the world, including the former colonial world and the critique articulated mainly from there, not to mention their role in universalizing this "Western" social science approach

What could better illustrate the global reign of social science thinking consisting of a battle among national and global knowledge bodies, than all these critical discourses about the world social sciences, may this be about global "in-equalities", "asymmetries", "academic dependence" or "hegemonic sciences", "knowledge flows", all considering the world social sciences as consisting of "national science communities" and their imagined regional alliances, to which social sciences are associated irrespective of what they say simple thanks to their nationality, all nationally constructed knowledge bodies, which according to the views of this opposition, opposing the battle among national knowledge bodies—heavily arguing about the distribution of weapons for this battle. The nationality of academics in all those endless bibliographical studies, critically tracing a ranking among national science communities, all their eagerness to identify the nationalities of academics, does not only not want to critique a science world that consists of isolated state thinking knowledge islands, nor does it want to oppose a science world that argues about the status national knowledge bodies hold, nor the idea, originally national science policies imposed into the social sciences interpreting discourses among social sciences across the national knowledge bodies as a battle about the status and prestige of national knowledge bodies hold against others, but rather shares this view of the national science policies and is heavily engaged in debating an "asymmetry" and the like of the weapons these nationally constructed creatures, only national science policies make from them, have within these battles, a view only these national science policies can originally create with their views on the social sciences as a means for their battles about real global power, a view this critical discourse without any hesitations shares, joins and further develops towards theories about an always—unfair—"global knowledge production". Social science thinking is critical thinking and therefore cannot resist to compliment the fact that social scientist are forced into this battle by the national science policies, making social scientist representatives of their nation states and members of such academics troops, to—as ever critically—articulate their committed concerns about the asymmetries of the weapons they are given by their political masters for this battle.

The progress of acknowledged knowledge

The global discourse about knowledge under the social sciences is about the question, which of the many nationally constructed knowledges are those globally acknowledged knowledges that rule global theorizing, accompanied by the above discourses, arguing about the means to rule theorizing and a critique that complains about the "unequal „distribution of "scientific power" in this battle.

However, though the question how acknowledged knowledge argues as knowledge, what the discourse is about that aims at becoming globally acknowledged knowledge framing theorizing across the world's social sciences, though this question has—at least implicitly—already been answered, the question, what finally qualifies a theory that replaces others as by a new meta-theory ruling from there on theorizing across the world, still should be discussed a bit more elaborately. How—to phrase it in Sato's words—to create "thin and fast spreading concepts", how to become a *globally* acknowledged "grand theory", what qualifies knowledge as world-wide acknowledged knowledge to become the new acknowledged knowledge, framing a new wave of theorizing in the social sciences, a theoretical "turn" in social sciences, or in other words: How does social science knowledge, how does acknowledged knowledge process?

This is not only a question, since the same T. Kuhn, so much appreciated by the very social sciences as a witness for the social science dogma, the impossibility of creating knowledge while discussing the knowledge achievements of the natural sciences, states in a remark about the social sciences that the social sciences have until today not reached any knowledge that could be considered as a "paradigm", hence, are also lacking any paradigm shift. What means in other words, that the social sciences do not only not own any knowledge, shared by the whole social sciences; it also implies nothing less but two other things: firstly, that the social sciences in their more than 200 years history of theorizing did not arrive at any secure knowledge and, secondly, that there is no progress of knowledge in the social sciences in the sense that their thinking increases the knowledge they have about the world's social.

And this is not all: Another account of 100 years of social science theorizing by another social science guru, E. Wallerstein, complements this disastrous account of social science thinking with an even more irritating and also telling account about the social sciences achievements, saying that social sciences until today cannot even precisely say what their subject matter is.[152]

Let aside the problematic of the notion "paradigm" and the above critiqued conclusions from this notion, what T. Kuhn is saying, is that unlike in the natural sciences the social sciences do not have any knowledge, one could consider as knowledge shared by the social sciences. Seemingly, this seems to be worth mentioning by a natural scientist, for social sciences, however, this is only an implication of their dogma, that any knowledge must be ever relative knowledge and thus only the consequence of the nature of any

---

[152] Wallerstein, I., (2001) *The Limits on Nineteenth-Century Paradigm, Unthinking Social Science,* Second Edition, Temple University Press, Philadelphia,

knowledge. For social science, therefore raising the question, how their knowledge progresses only indicates a violation of their dogma that there is no secure knowledge, hence there is no such thing like the progress of knowledge—which is, however, also not true.

Such observations about the not existing knowledge progress in the social sciences, no social scientists seemingly finds worth to notice or to mention, thanks to the anti-scientific skepticism of social scientists, may they be natural or social scientists, when they reflect about science, they find such an account of 200 years of thinking about the world only most natural and for the very same reason celebrate the theories of T. Kuhn for his false theory about the progress of knowledge in the natural sciences, serving this anti-scientism, they present as *the* one and only secure insight they have, an insight into what science must be like, not in order to omit this odd job of useless theorizing, but to sacrifice their lives for coping with this hard mission of a professional thinker, providing approximations to understand an essentially un-understandable world.

Social science knowledge does progress, it does create new theories, critiquing their predecessors, may this be in the odd way as discussed above as a critique which accuses theories they replace with the non-critique that they are not saying what the new theory says.

While the history of natural science is indeed a history of correcting false insights, a correction the epistemological debates among social sciences, frivol enough interpret as a prove that there is no correct knowledge, social science do not know such history, progressing with the knowledge they have by correcting existing or adding new knowledge they did not have so far, but their knowledge does though progress, not with knowledge about social phenomena that did not exist so far, but with new theories about the same thing about which they already own theories.

How they introduce new theories replacing knowledge they have with new knowledge without critiquing the existing knowledge has been shown above. How social sciences create new theories staying within the existing grand theories, is the ordinary business of the social science theory production. There are, however, undoubtedly shifts in social science theorizing, new theories appear and the question about the progress of knowledge in the social sciences is then, how new theories progress beyond the existing knowledge without eliminating their errors of their old theories.

Hence, the question remains, if social sciences do not replace old theories by a new theories, because they discover that the old theory did not or did only falsely explain its object of thinking; or, to phrase in in other words, if the enlightening justification for the need for a new theory is—as in the above case of Skinner—the tautological critique that it is the mistake of the old theory that it does not do what the new theory suggests, how does this way of theorizing discover the need to progress from the knowledge they have towards a new theory, if it cannot find the reason to progress from this knowledge towards new knowledge in the knowledge they *have*? How does knowledge become acknowledged knowledge and why would then this knowledge need to be again replaced by new knowledge? What is the nature of such ephemerally progressing ephemerally true knowledge?

Social science have an answer on this question, however this answer is also obviously wrong: Social sciences ever justify the necessity that social thought must ever be ephemerally true knowledge due to the nature of the social, due an ever "change" of the social. This is a false thought, since, no doubt, social life creates phenomena which are new; however, the conclusion that therefore knowledge about a no longer existing social phenomenon is no longer knowledge and calls for new knowledge, is a false conclusion. Reasoning the need to progress with the creation of new theories with the changing social reality and to consider the developing reality, which would require rather adding new elements to an existing theory, instead disqualifying an entire theory due to any shift the social reality made, just as if every change was a revolution of the essentials of a society, reasoning the need to progress with the creation of new theories with the changing social reality does not only reveal an a-historic concept about how knowledge really progresses with the changing reality, social sciences ever phrase the need to progress with their knowledge, because they consider the knowledge they have as being *"outdated"*, theories from yesterday. And this notion of critiquing knowledge as "outdated" to argue for new knowledge contains some hints about how this knowledge really does progress.

The case, that in fact an alternative society system, let's say, the society system "socialism", has disappeared from the world, does not allow to conclude that theories about this society system are outdated, because the object of thinking does no longer exists. To conclude from the fact that a phenomenon does no longer exist that a theory about this phenomenon is no longer true is a false conclusion. It might be the case, though not for historians, that knowledge about any no longer existing or substantially changed realities diminishes the scientific interest in theories about them. Presenting however theories about a no longer existing reality as being "outdated" theories which causes the need for new theories is as false as indicating why and how social science knowledge changes. If it was true that it is the changing reality that causes the change of social science theories, social sciences theories would have no history and, hence, no progress. Indeed, the critique of knowledge, which social sciences use to argue for the need of new knowledge is that knowledge is "outdated" is as false as instructive to answer the question how knowledge proceeds that must be ever updated, ever motivated to *catch up with the "change."*

"Outdated" is a theory not because the reality has changed, but because social science thinking theorizes through an "updated" model theory through which they construct their knowledge. Another tautology such as the notion of a paradigm? Indeed, while the notion of a paradigm applied to natural science, that progresses via the elimination of false theories towards correct theories, mystifies their progress of knowledge towards any tautologically constructed circumstances initiating a paradigm shift, the knowledge progress in the social sciences is no progress of knowledge in the sense of eliminating false knowledge, but—as we will see along the two below discussed examples—*a reflex of social sciences on a changing interpretation of the essentially same reality,* a tribute social science thinking pays to changing circumstances, exactly the very mystic interplay of any mystic circumstances and theorizing, Kuhn coined with the notion of a paradigm. However, as we will see the mystic interplay of theorizing and a changing reality which only changes its own interpretation of itself, is not at all as mystic and the tautological logic of how social sciences proceed is the logic of knowledge that *progresses and says something new without progressing from false to right* knowledge.

Again, the need for new knowledge in the social sciences never critiques the existing knowledge as knowledge that needs to be amended, knowledge that must revise theories or add new insights to the knowledge they have; social sciences rather detect the need for new knowledge with the argument, that any new explanatory model theories are needed, even for phenomena that already exist or existed, as Skinner's example documents with his resonations about the need for a new psychological theory he coins "behaviorism". Social sciences, in fact, detect the need for progressing towards new modelling theories, because they detect a new model for seeing *the old and the new social world;* updated knowledge is knowledge, that is constructed through such new explanatory model theories, as the one Skinner has introduced or the one such as Habermas presents his theory about "The re-construction of Historical Materialism" ,as a theory that allows to understand all the social sciences meta-theories he declares as predecessors of his theory. In psychological theorizing, the shift towards theories about the new notion of "cognition" is no theory about any new phenomenon, but a new approach to psychological thinking. The concept of "modernity" in sociological thinking, though it does play with a historizing category is no theory about any new historical period, but a new model to theorize about a historical period, which before was coined by more Marxist theories as capitalism, implying a whole set of categories through which the same reality was explained as the one which has then become a case for theorizing about "modernity". Econometric theories are no theories about any contemporary economic phenomenon, but a contemporary approach to economic thinking, an approach allowing a French economist to write a Nobel prized book consisting of tables with figures about throughout almost the whole recent history of capitalism, figures to illustrate a new theory phrased in one sentence. Needless to repeat, all these theories, representing a progress in social science theorizing, present their new model theories with the critique of their preceding models that they are "outdated" because they did not do what the new models do.

Social science thinking obviously progresses by creating updated model theories, constituting a new wave of theorizing constructed through these *updated views interpreting the world,* a wave called "mainstream" thought, interpreting the view of social science theories through those new theories which are serving as new grand theories, through which theorizing from then on is ruled, those grand theories all the above mentioned criticists of the "Western" social science envy the "Western" social sciences for that it is "Western" social science who are ever setting them as theoretical standards for "global" social science theorizing.

The question, then, how do social science progress, is therefore more precisely the question, how does such a new model theory gain the status of a new model theory for the social sciences as a whole or for theorizing in disciplinary thinking using this metatheory as theoretical framework theory framing their pre-supposed way of theorizing.

The option, that any new knowledge progresses in the social sciences as replacing preceding knowledge, because it is false knowledge, one can exclude as an option within the social sciences approach to social thought, since this provides a critique of the mistakes of a critiqued theories, but critiquing theories as false theories is no option in a social sciences which knows only theories which are always relatively true theories, relatively true dependent on the theoretical framework and the methodologies through which they theorize. Hence, one must look at how these metatheories change without changing due to a critique of the mistakes of the theories their replace.

And additionally, focusing in this book on social thought in the social sciences about the world's social, may be shorter phrased global social science thought, the question is how a theory becomes one of the acknowledged true global social science theories, one of the globally shared ways to interpret the social framing global social science theorizing?

How to create a globally shared truth ruling global theorizing

The easiest way to answer the tricky question how scientific knowledge, in which the objective matter of truth is the subjective matter of a shared truth, is progressing is to look at how two examples of theories which are examples of knowledge that advanced theorizing towards thinking through these theories, how these theories became shared true theories guiding theorizing across the social science world.

The two following examples of theories, one from economics, the theory about the "knowledge based economy", and the second one from sociology, the theory coined "risk society", will answer the question how to create a theory that provides globally acknowledged knowledge and, as such, rules global social science theorizing.

Example 1: The theory of a "knowledge based economy"—
The alternative imperial agenda of the Europeans

How "mainstream" thinking works, from where it receives its impulses and inspirations and how it transforms thinking into a new mainstream theories, originating from thinking about the economy and from their enters economic theory to then from there invades the whole contemporary social science thinking without any critique of any previous economic theory, a transformation of social science theorizing that results in adjusting the conceptual apparatus across all disciplines to this new ruling theory, this can be demonstrated with the example of the theory known under the notion of a "knowledge based economy".

To reveal the result: required in the first place is jettisoning of all scientific scruples to be able to present all the irrationalism of a market economy as an economy guided by nothing less but by knowledge—that very knowledge that never really knows anything. And to present a market economy as an economy guided by knowledge, therefore called a knowledge-based economy, it is secondly required to not have any knowledge about the market economy. Instead, one must have above all an intimate sense of the universalism of the agenda of imperial nation states, here the European Union to become a universal meta-theory for theorizing.

There is a notion in economic theories, that ever challenged economic theorizing because it conceptually undermines the mission economic thinking attributes to the economy, a mission committed to present the market economy as providing the society with goods and services the society needs, a challenge that was waiting for the chance for economic theorizing to create an updated version of economic theories and to go beyond the slightly restrictive notion of the economy as ever fighting against "scarcity". The slightly negative notion of scarcity conflicts with the idea of a market economy providing people with goods since it reminds too much of a real scarcity some people encounter especially in the notion of "capitalism" and, to make it worse, even more in the notion of global capitalism. Replacing this negative image by a more positive view on the global market economy is the mission of a theory coining global capitalism as a "knowledge based economy", an economy guided by nothing else but such a honorable thing like knowledge, thus counter coining the so-called notion of a "neo-liberalism" US economists advocate, with a truly European alternative way of coining an imperial global agenda as an economy guided by knowledge—an agenda with the very same substance, now named with what the enlightened European capitalist after all really is: philosophically inspired business people aiming at competing with the US economy about something where they are unbeatable by those US cowboys—knowledge.

To start with the impact this theory has both on theorizing as on social practices it might be helpful to sketch the success story of this theory, before discussing its thought in more details, which is worth doing to understand the cognitive steps economic thinking makes to gain the status of a global mainstream theory, setting the categories for theorizing, across disciplines and across the world.

The success story of the theory of a knowledge based economy revolutionizing social practices is, in fact, is remarkable: This theory has not only provided the Zeitgeist with a whole terminology towards a substantial shift of the national policy agendas, it did this for the policy agendas across the whole world, at least there, where the Europeans managed to get to play an influential imperial role. The policy shift for which the knowledge-based economy theory has provided a new view on the world was not at all only a new agenda only for the sphere of economics but an agenda across all policy spheres, a policy shift which was about nothing less but a global reform of the world's nation states agendas transforming all policy domains and with this the whole society into a service for the global battle about the growth of capital, preferably presented as caring about jobs for people in the global battle about the requirements of a knowledge based economy. And to repeat this, it provided not only the Zeitgeist for this policy shift, elsewhere called "neo-liberalism", in one or a group of countries but throughout the policy agendas of the entire world of nation states and in the entire world the agendas of nation states, both for their internal affairs and for their external imperial agendas. Redefining the whole policy agenda towards a society functionalized for serving as a resource for attracting global capital was and is the policy rationale for which the theory about a knowledge-based economy has provided the theoretical ideas and the set of categories under which these reforms were labelled, conceptualized and executed.

The realm of science was on other of the major construction areas which was substantially re-constructed in the name of this theory, both institutionally and categorically.

The "Bologna process" phrased for the academia to create a "Europe of knowledge" as nothing less than the policy mission for the new millennium:

> "A Europe of Knowledge is now widely recognized as an irreplaceable factor for social and human growth and as an indispensable component to consolidate and enrich the European citizenship, capable of giving its citizens the necessary competences to face the challenges of the new millennium..."[153]

Under the auspices of the knowledge based economy Higher Education firstly within Europe and from there across the world, by creating a global Higher Education market challenging the so far globally reigning US academia, the world's academia was revolutionized towards a means of and part of a global market economy, replacing all the previous humanistic ideas and ideals about academia through the spirit of making science a business.

The theory of a knowledge-based economy has not only provided the theory for revolutionizing social practices across the world, but has also revolutionized social science theories. Thanks to this theory social science thinking across all disciplines and again across all the global state science islands has been invaded by economic categories replacing the whole set of categories that originated from idealisms of European humanism.

To mention just two examples from the realm of educational thinking and from thinking about science:

While education, learning, was by then presented as serving the development of the individual, the notion of *competences* redirected the acquisition of knowledge to adjusting the cognitive abilities of humans towards a function for the technical needs of the production process, the knowledge based economy theory calls this expropriation of knowledge, knowledge *"tied to the people"*[154],

---

[153] *The Bologna Declaration of 19 June 1999*, http://www.bologna-berlin2 003.de/pdf/bologna_declaration.pdf

[154] B.-A. Lundvall, & S. Borras, (1999) *The Globalizing Learning Economy: Implications for Innovation Policy,* European Communities, Luxembourg: Office for official publications of the European Commission, p 60

from the people towards a functional service for the capital, *"adapting constantly to the new demands and conditions of the economy"*[155].

The social sciences adapted their way of theorizing to the new demands and conditions of theorizing in the category world of the knowledge-based economy theory. Categories from the realm of economic thinking once reserved for economic theories have invaded the categories across all disciplines and since then theorizing about the world, the creation of knowledge has become a "knowledge production", knowledge a means for "credits" and "rankings", universities became "flagships" and discourses among academics became "knowledge diffusions" wherein knowledge since then is "exchanged". And so on and on.

In fact, a success story, that managed to set models for theorizing and categories within the world's knowledge islands and across the categories of disciplinary thinking.

Two questions still need to be answered: What is the theory of the knowledge-based economy and what is the reason for its success, both in re-structuring and re-phrasing policies and in rephrasing the whole set of categories in the social sciences and there across all disciplines?

To start again with the result: The argumentative structure of the knowledge based economy can scientifically only be called scandalous, also from what social sciences would consider as "good" science. However, in social sciences, where false a right thought does not matter, this scandalous way of theorizing was not recognized as scandalous, since the scientific quality of this theory is in principle not that different from any other theory, though, admittedly, an example for a kind of frivol free floating spinning of ideas, decorated with some philosophical categories, the economist, fantasizing about knowledge towards a knowledge-based economy do not need to understand to use them for constructing their new meta-theory setting the grounds for a new social science fashion.

What is according to this theory a knowledge-based economy?

> 'A knowledge driven economy is one in which the generation and the exploitation of knowledge has come to play the predominant part in the creation of wealth......It is argued that the knowledge economy is different from the traditional industrial economy because knowledge is fundamentally different from

---

[155] Ibid p 61

other commodities, and that these differences, consequently, have fundamental implications both for public policy and for the mode of organization of a knowledge economy.' Joseph Stiglitz (1999), for instance, suggests, the 'movement to the knowledge economy necessitates a rethinking of economic fundamentals because, he maintains, knowledge is different from other goods in that it shares many of the properties of a global public good.'[156]

How does economic thinking rethink "the traditional industrial economy" to reason the occurrence of a knowledge driven economy?

Step 1 to rethink the economic fundamentals the knowledge-based economy theory starts with reflecting about concepts of knowledge and to does this by referring to the Polanyian concept of knowledge, distinguishing between tacit and codified knowledge.[157]

> 'When discussing the role of knowledge and knowledge production in economic activity it is important to distinguish between tacit and codified knowledge.'[158]

To distinguish both concepts of knowledge is important for rethinking the economic fundamentals towards a knowledge-based economy, because both concepts are attributed a nature by the rethinking economists, which has not only nothing to do with the concept of either tacit or codified knowledge.

> 'Codified knowledge can normally be transferred over long distances and across organisational boundaries'[159]

---

[156] M. Peters, (2003) *Knowledge Networks, Innovation and Development: Education After Modernization.* Paper presented on the EURONET workshop, Stirling October (not published)

[157] The fact that Polanyi theorizes about a theory of thinking, in which this distinction between tacit and codified knowledge actually only rarely occurs, because it is of a very minor importance in his theory, is the usual ignorance of social science theorizing that uses any knowledge that fits into their theory modelling, no matter what the original theory really is about. It was actually education theories which interpreted Polany's theory about thinking as a theory about learning long time before economist used this distinction for their purposes to construct their economic theory about the knowledge-based society. A pinch of philosophy always adds the taste of science to the most vulgar way of constructing theories as if this was about adding a bit of this and that as in a recipe for cooking a theory. Polanyi, M., (1958) Personal Knowledge. Towards a postcritical philosophy, Chicago, University Press

[158] B.-A. Lundvall, & S. Borras, (1999) *The Globalizing Learning Economy:...* p 31

[159] Ibid

No matter if this what the theory about tacit and codified knowledge means is the case or not, according to the theory about codified and tacit knowledge both knowledge types are distinguished by their differences acquiring knowledge, not by a difference being "*transferred over long distances and across organisational boundaries*" or not. The knowledge economists experts use a kind of most plump mistaken interpretation of the word "transferability" and identify the transfer of knowledge through learning, the cognitive operations of a knowledge "transfer" as being the same as the "transfer" economists can only ever think of, the economic transfer of commodities, which is the buying and selling of goods, thus, for economist no problem, can also be used for the "transfer" of knowledge in this economic sense. Since codified knowledge suits into the world of economic thinking, they feel they should say that they like it:

> 'Codification is an important process for economic activity and development for four main reasons. Firstly, codification reduces some of the costs of the process of knowledge acquisition and technology dissemination.
>
> Secondly, through codification, knowledge is acquiring more and more the properties of a commodity.'[160]

Codified knowledge gains more sympathy of the knowledge economists since it acquires "*more and more the properties of a commodity*"... which means it can be "*bought from the shelf*"[161]. Knowledge-based economists like codified knowledge, it gives them the feeling while discussing theories about knowledge as if they were in a good old shop.

Step 2: However, once the knowledge economists have translated the different characteristics of the two types of knowledge distinguishing modes of cognition into their distinctive abilities to be transferred as a commodity, that is bought and sold, they come to the conclusion that the whole distinction between both, which was so important to make, is entirely irrelevant.

---

[160] Ibid, p 32
[161] Ibid, p 46

> 'In fact, the clear distinction made above between tacit and codified knowledge may be misleading in some regards'[162]

The logic of thinking in modelled thought is to carry the thinkers to the result he wants to arrive at and to give this intention the image of any plausibility with a taste of a philosophical knowledge background. Forget about all this, it is misleading since it does not lead these thinkers—where? Where they want to go to. Since the interpretation of transferring knowledge as the economic act of buying and selling does not help the knowledge economist thinker to reason the fundamental role he wants to attribute to knowledge as a whole they simply discards all his previous thought stressing the importance of this distinction and call it now "misleading", since it does not lead him where he what he is up to. Let's forget about it. End of step 2.

Step 3: Translating codified and tacit knowledge into features of knowledge that attribute to knowledge the features of a commodity allowing to purchase knowledge was a good idea, but reminds him of a unresolved problem he made us believe he got rid of with his plump identification of transferring knowledge: Buying knowledge means to own this knowledge, this is fine. However, buying knowledge does not mean to possess knowledge intellectually, buying knowledge is not acquiring knowledge, the transfer of commodities is not the same as having the expertise of a knowledge holder, the knowledge an "expert" has.

> 'Most codes relating to science, technology and innovation can only be decoded by experts who have already invested heavily in learning the codes'[163]

What a mess, this is what economists did not invest in, they may have dreamt about codified knowledge that this knowledge can be bought from the shelf and now frustrated must acknowledge, that this, buying knowledge is an illusion because it does not mean that any knowledge that can be bought means the buyer gains knowledge by buying knowledge, the expertise still only the expert has. This is in fact true: knowledge cannot be bought and therefore the whole distinction is—useless.

Thus, he concludes:

---

[162] Ibid, p 33
[163] Ibid, p 33

'So far it is only in science fiction that mad criminals manage to get physical control of the brains of eminent scientists.'[164]

Knowledge economists are no mad criminals, but "so far" share their very dream: Getting "control" over knowledge by buying knowledge. A dream, they share with mad criminals; however, since they are eminent scientists, they are not that mad to not notice that buying knowledge was a nice idea, but that it is a fiction *"to get physical control of the brains of eminent scientists"*.
However, they know a way how to finally *"get physical control of the brains"* of eminent scientists' and the first step to arrive at a control, not a physical one, but a control, is to forget everything they said so far about the before important distinction of the two knowledge types. And we will see soon, why it is only "so far" a dream of a not that mad economist, a dream the economist still dreams of, and not only that, a dream he obviously already knows how to really achieve what only mad criminals are only able to dream of. The economist knows a way how to make this dream become true.

Step 4 is to not only forget all that has been said so far about tacit and codified knowledge and to go back to the fundamentals of economic thinking where knowledge economists feels at home. To reason the creation of the theorem of a knowledge-based economy that reflects on the new role of knowledge for the market economy, coining the market economy a knowledge based economy—the economists go on to prove now nothing else but—the incompatibility between knowledge and the fundamentals of the market economy:

"Knowledge ....defies some fundamental economic principles.'[165]

"The problem with using the market failure concept in the context of the learning economy is that almost all aspects of knowledge creation and learning are characterised by market failure... These characteristics make information/codified knowledge a very peculiar commodity and all transactions involving it will be characterised by elements of market failure. Tacit knowledge

---

[164] Ibid, p 46
[165] OECD, (Organization for Economic Cooperation and Development) *The Knowledge Based Economy*, Paris: Organization For Economic Organisation and Development, 1996, p 11

is a plain market failure in the sense that it cannot, as such, be transacted in the market."[166]

This, concluding that knowledge is in contradiction with fundamental economic principles of the market economy, does not disprove the theory of the knowledge-based economy as the latest version of the market economy, but *is* a proof—because the knowledge economists wanted to arrive at this dis-proof, a disproof he needs to prove the need for the intervention of politics to make his dream reality:

> 'This being the case, we can conclude that the market failure is not a useful concept in the learning economy.'[167]

Again, the logic of economic thinking is to force thought towards what the thinker wants to think. If knowledge is a market failure, then the conclusion is not that knowledge cannot be a fundament of the market economy, but that the market failure proves that it was a wrong concept to be applied to knowledge, to prove what he is determined to prove. This brings him to step 5.

Step 5: In fact, it is exactly this failure that knowledge does not obey the rules of the market that proves that the new version of the market economy must be a knowledge-based economy. How come?

> 'Since this applies wherever innovation matters it gives little help in locating a need for policies.' [168]

Proving knowledge as market failure calls for a need for policies. His proof of knowledge as a market failure *"gives little help in locating a need for policies.'* The knowledge economist straight forwardly was determined to arrive at proving knowledge as a market failure because it allows him to invite politics to practically correct the recalcitrant nature of knowledge and to make the dream of a knowledge-based economy come true, mad economists "so far" only dreamt of.

What the recalcitrant nature of knowledge is, is that knowledge defies to get under the control, a control mad criminals only dream

---

[166] B.-A. Lundvall, & S. Borras, (1999) *The Globalizing Learning Economy:*...p 49
[167] Ibid
[168] Ibid

of unlike eminent economist thinkers, who know that calling the help of the political power will make this dream reality.

To disclose the 'locked in knowledge'[169] in order to guarantee its economic functionalities is the mission, economists know must be delegated to politics, which has the power to force the market failure to become a market success and to thus also to do the economist, who called for a little help from the political power, the little favor, to thus allow him to then rightly coin his new economic model for theorizing a knowledge-based economy.

> 'Learning and knowledge are tied to people, and if people cannot keep pace, there is little point in having access to advanced machinery or advanced computer programmes. ...The need to stimulate investment in human resources and organisational change at the firm level has become more widely recognised by policy makers.'[170]

Knowledge politics have to use their political power to untie learning and knowledge from people and to make it a public good that is to domesticate people as "human resources" and their brains a means for the global competition of capital.

To recapitulate: What are the theoretical steps to arrive at a theory proving that the essential of the economy is knowledge, therefore called knowledge based economy?

The eclecticism with which economists craft their theories, is breath taking, though it is entirely normal for social science thinking to construct thought squeezed into the theoretical framework of presupposed thinking their categories and thought must serve and obey:

Firstly, crafting a theory about a knowledge-based economy inevitable calls for some reflections on knowledge. To do this creators of a new economic theory, that conquers the world of theorizing, borrow two categories from a philosopher, tacit and codified knowledge, and, ignorant of what these categories mean, they very free handedly interpret both the transferability of tacit and codified through learning as the same as transferring knowledge through buying. Then, after discussing this distinction, they discard this distinction to say, that it does not make sense to make this distinction, because neither of those categories supports a theory that

---

[169] D. Rutherford, *Dictionary of Economics,* 1992, London, Routledge, p 272
[170] OECD, The Knowledge Based Economy,.... p 60

knowledge can be transferred, which is bought, because neither transferring nor buying allows to get command over any knowledge, only experts have. Then, after confessing that the whole debate about knowledge was a useless dead end road, they, thirdly, raise the question, if knowledge, the essential of their deduction of a knowledge-based economy as the new version of a market economy, is at all a commodity, something the obeys the rules of a market economy and arrive at the result, that knowledge defies all the essentials of a market economy. This, stating that knowledge does not coincide with what a market economy and rather defies its basis element, a commodity, does not terminate their reflections on developing a theory of a new type of market economy founded on knowledge, but is, fourthly, the reason they present to call for political power to force knowledge to be a commodity that it defies to be. Thus, the deduction of a theory is the result of the intervention of political power they call to make a theory true they theoretically cannot prove. A true master piece of social science theorizing and a masterpiece of how social science thinking creates theories, providing a set of presuppositions, categories for social science theorizing across the world and across all disciplines. Scandalous? Certainly a bit more vulgar, but not really that much different from the above discussed ways how disciplinary thinking across all disciplines argue to found the categorical basis for their disciplinary category systems.

The mission of economic theorizing is completed, the unresolvable problem to get knowledge under the control of economic needs of global capital is solved with the little help of making knowledge a public good under the control of nation state politics presented under the label of a society that aims at knowledge. A truly smart way of theorizing: Since the theorist failed to prove that knowledge is an economic good and that knowledge even defies the rules of the market, these social scientists conclude that the intervention of the political power is needed to force knowledge under the rules of the market and to serve the needs of the market its nature defies.

Under the noble image of a society governed by knowledge this theory not only provides the theoretical titles for the intervention of politics restructuring nation state policies across the world towards and for the needs of global capital but calls for this intervention to create with the forces of a nation state power, what the scientific thinker proved was—otherwise—not possible. Only "mad criminals" and smart economists know why this is a knowledge-based society, because the nation state must force knowledge to obey its political agenda. The economist has no doubt that the nation state power will do what he suggests. Or is it rather the other way round?

Thus, not only the notion of a knowledge economy since then has become the noble title for the European version of the recent agenda of imperial nation states, elsewhere with exactly the same political rationale there coined as neo-liberalism, in Europe and from Europe across the world, now established with all kind of noble titles derived from the notion of a knowledge-based economy decorating an agenda that makes the world's social a means of global capital and its supervisors, the imperial nation states.

Coining a society that subordinates thought and the acquirement of knowledge as the continuous adaptation of alien knowledge under its use for alien objectives, in short calling knowledge "competencies", calling this a knowledge-based economy is certainly impudent, but it is this impudence that illustrates how social sciences thinking works and what it is that qualifies social science theories as theories guiding the creation of social thought in the social sciences. As in this case of a theory coined as a knowledge-based economy, it is a theory that guides global social science theorizing, *because it succeeded in re-phrasing the agenda of imperial nation states as an agenda aiming at a society of knowledgeable people* and it is this, *presenting an imperial agenda as a service for mankind's need for knowledge*, that makes this theory a global meta theory guiding global social science theorizing.

It is not at all only in economics that the theory of a knowledge based society invaded. It invaded social science thinking across all disciplines with their categories framing the creation of theories. The economization of the categorical apparatus of the social science disciplines, that is the penetration of disciplinary thinking with categories originating from economic thinking and thus transforming social science theorizing towards thinking that interprets any social activity with economic categories, such as "competencies", "knowledge production" and so on and on, illustrates the successful invasion of a theory into social science theorizing that presents *the subordination of all spheres of social life under the needs of the global competitions of capital as the rise of a society aiming at knowledge;* the appropriation of knowledge by and for the needs of the globalized capital presented as a society of thinkers.

And a new meta theory was born guiding social thought across the world. This is the theory that has since then framed and frames theorizing about the social, not at all only within economic theories. This is the theory that gave rise to all the new categories across all social science disciplines and this is the theory that revolutionized social science theorizing across the world and across all major social science disciplines. And: This is *the* theory that paved the cognitive ways for social science theorizing in the era of "globalization". A new meta-theory for theorizing was created, an example of progressing social science knowledge.

Example 2: Rien ne vas plus: The "risk society"—a repository for critical social science thinking about the apocalypse of the "unworld"

The theory about the "risk society" [171] rightly is one of the most quoted social science theories in the world, since it represents both the social science mode of theorizing as well as the kind of social thought this way of theorizing creates as scientific thought about the nature of the world's social: It is the nature of the "risk society" that makes thinking about the social an impossible endeavor and it is the nature of the risk society that the social world is the materialization of the disappointment of social sciences that it is not what social sciences think it supposedly should be, in short the social science world is the "un-world".

It is certainly the case, that the explosion of a nuclear power station in the Ukraine made his book very suddenly very popular, also among non-academic readers. However, one should leave the malice his book has received to the political conservatives, calling his theory a "theory of a freehand sailor", and to those academics, who only envy him for the success they vainly try to achieve with exactly the same theoretical endeavors. The history of social science, namely sociology since Marx, is full of such attempts to prove his theory as outdated and that societies, at least in the imperial world,

---

[171] U. Beck, *Risikogesellschaft. Auf den Weg in die Moderne,* Frankfurt 1986, Suhrkamp

are no longer a "class society" as the creation of a "middle class", which, though it is still called a "class", is proving that the society is no longer divided so simply into poor people here and rich there, since there are people which are not that poor, if one just counts the statistic figures in a more sophisticated way, just as the new economic superstar Piketty does. A ghost still goes around in Europe, it is the ghost, that the world's societies are still capitalist class societies and social scientists are ever obsessed by the mission to prove that this is not the case, and that theories like Marx's, might have been somewhat true in the society people like Beck prefer to circumscribe more technically as "industrialization" but which are an old "outdated" story today.

So U. Beck: He wrote a book about the "The risk society", a notion that also argues against the existence of a class society and suggests instead his theory of a "risk society", a theory that symbolizes the nature of social science theorizing, both epistemologically regarding the concept of social science theorizing as the contents of theories this mode of theorizing creates.

This book about the "risk society" has become a theory framing social thought across the world, and the unexpected promotion it received by the blasting nuclear power station, might be responsible for its popularity among ordinary readers, however this does not explain why his book is one of the most cited books among academics across the world. In fact, Beck only did what numerous social scientists tried before him and that is to create a theory about the society that discusses a social world beyond the concept of class society and did this with very successfully.

As a typical sociologist, discussing the class society as a *possible, though outdated* explanatory model and what typical sociologists, who want to take part in mainstream sociological thinking, see comes after the class society, the concept of "modernity" as a new model for thinking about the society, in Beck's scenarios he prefers to coin with the notion of a "risk society", it is not the concern of all these most sociological constructs, to analyze what the existing society is, but what could be an "updated" theoretical model for theorizing about the society or the social in general. It is not the concern of such meta-reflections to deny the relations between poverty and wealth; for social science thinking, in particular sociological thinking, it is not that they are bothered by the fact that people work for their entire live and end up as cases for the social security system. What bothers sociological thinking about the class theory as any other modelling of society is that the class theory is a theory, an explanatory tool framing thinking that *divides* the society into anything, Marxists into classes and it is this idea of dividing a society, which for sociological thinking is an explanatory approach that is a "risk" for the one and only sociological concern they ever discuss, that is their concern to think about anything social, which is failing to create theories, a way of theoretically modelling the society that presents the society as to *unify* the inhabitants of this society and to serve their inhabitant's desire for guidance, orientations, common values, in short theories, framing social science thinking through the mission of society, Germans called and call "*Ordnung*". Not any particular "*Ordnung*", the nation state might be just one model for *Ordnung*, but at least any *Ordnung* and the nation state—who wants to object this insight into the "facts" is one possible *Ordnung*, not an ideal one, one, as sociological thinking is ever warning, that mostly fails to create this *Ordnung*, but is one realization of these sociological ideas sociologists otherwise like to call more abstractly "structure", "system" and the like. Anyway, it is any of these categories, idealizing the nation state as providing any kind of *Ordnung* that must above all guide social science theorizing—unlike the dividing idea in the notion of a class society.

To provide a theoretical framework that serves to interpret the social as always seeking this "*Ordnung*", a structure, a systems etc., social science theories must do two things: Firstly, present the social reality as a theoretically not-to-decipher darkness, that, secondly, needs nothing more but the insights of social sciences, namely sociological thinking, penetrating this impenetrable darkness to detect that man needs above all the social sciences thinking that provide insights about ways towards what man above all needs, systems, *Ordnung* and such things creating structures and *Ordnung*, interpretations of the society providing guidance for those in the darkness. It is not very demanding to notice a very religious mysticism that guides a concept of theorizing that ever proves that it is the nature of the social to mystify what it is and that humans need help to find their ways through this darkness, a darkness into which social sciences theorizing has transformed the social world to then appear as a dues ex machine enlightening those humans who are blindly wandering around in the darkness—until they find a sociological theory, a book that shows the ways out of the darkness by telling them that they live in the very ordinary darkness—of a risk society, proving that sociologist found millions of bad reasons in the risk society to call for what the lost citizen is longing for: any "Ordnungssystem".

The way sociological thinking persuades the world that the world needs nothing more but Ordnung, is always the same: It is to passionately describe how much people suffer under what sociologists diagnose as—exactly—, the darkness they cannot not only not escape from, the absence of *Ordnung* that provides orientation, but, worse than this, a darkness that is hiding that it is dark, a darkness ordinary people seeking for *Ordnung* even cannot see, sociological thinking, thanks god, can decipher[172]:

> "Der Unmittelbarkeit persönlich und sozial erlebten Elends steht heute die Unbegreiflichkeit von Zivilisationsgefährdungen gegenüber, die erst im verwissenschaftlichten Wissen bewußt werden und nicht direkt auf Primärerfahrungen zu beziehen sind."[173]

> "The immediacy of personally and socially experienced misery today faces the incomprehensibility of the civilisation risks, which only become conscious in scientific knowledge and are not relatable to primary experiences." (Own translation)

With this striking logic Becks theory about the impenetrability of the mystic darkness the explosion of a nuclear bomb has the nasty ability to hide that it explodes and remains in the darkness for

---

[172] The following three quotations from Beck's book are presented in the German original and complimented with my own English translation. Firstly, it is worth it to understand the original way of phrasing his thought, since it illustrates Becks manor, making the reader sharing his shiver. Secondly, unlike the German original, the official English edition not only loses this original spirit of creating through scientific theories the shivering and head shaking recipient Beck likes to create. The original English translation also does not allow to grasp the philosophical superelevations of Beck's appreciations of horror scenarios towards scientific thought thanks to a use of a very German philosophical terminology that gives his work, which substantially is not different from the tradition of German fairy tales for children, also presenting the world of adults as a dark threat. Boiled down to a more direct, though certainly a much less professional translation, both elements of his particular style of writing, become much more visible, than they could in the scientifically polished terminology in the official English version of his book.

[173] Ibid, p 68

those it kills and is therefore more than only once incomprehensible, it is only comprehensible when it is too late:

> "Der Effekt ist aber erst da, wenn er da ist, und dann ist er nicht mehr da, weil nichts mehr da ist. Diese apokalyptische Drohung hinterläßt also keine greifbaren Spuren im Jetzt ihrer Drohung."[174]
>
> "The effect is however only there, once it is there, and then it is no longer there, because there nothing any more. This apocalyptic threat hence does not leave any comprehensible roots in the threat of its now." (Own translation)

What an exquisite logic: Any catastrophe is firstly and above all, to not being visible as catastrophes, except for sound social science thinkers who are able to trace the roots of the threat of the nuclear bomb before it explodes and write a book distancing themselves from the stupid rest, who does not see all the apocalyptic things happening; this rest remains in the darkness of their lacking experiences with such explosions and thus cannot see them as a threat, a logic of the lonesome social scientist among all the blind people is a logic, which does not make this social scientist less dull than he thinks he is.

Then, secondly, and—easily to conclude from what the opposite of a catastrophe is—is a non-catastrophe, which is *Ordnung* and thus all catastrophes are for this way of apocalyptic thinking ever anyway essentially the same and do not need to be thought about other than finding them all as—catastrophes. Under the headline of "living on the volcano"[175] Beck, just as some grandmothers and—by the way—the Bible enjoys to do, without any reflexive scruples enumerates earthquakes, wars, car accidents, dictatorships, diseases, the hypocrisy of science, rationalism, in short whatever comes to a mind, that presents all this as a deeply disastrous world lacking insights towards the need for *Ordnung*, only—thanks god—the sociological thinker depicts for us dull people.

---

[174] Ibid, p 50
[175] Ibid, p 78

"...einer großen Bevölkerungsgruppe stehen heute, mit oder ohne Absicht, durch Unfall oder Katastrophen, im Frieden oder Krieg Verheerungen und Zerstörungen ins Haus, vor denen unsere Sprache versagt,...Es handelt sich um das absolute und unbegrenzte NICHT, das hier droht, das "Un" schlechthin, unvorstellbar, unbegreiflich, un-, un-, un-."[176]

"... major parts of the people are facing today, may this be their own intention or not, due to hazards or catastrophes, during peace and war, distortions and destructions, our language is no longer able to grasp,.... This is about the absolute and unlimited Nothingness, that is threating, the "Un" as such, unimaginable, incomprehensible, un-,un-,un." (Own translation)

And it was and is this way of thinking presenting the whole world as a dooming threat, it is this very Christian idea, painting a world of pain, chaos and dramas that was and is also appreciated and applied by academics, namely in the 1980s, and to our day, who use their lack of interest in understanding any of these "catastrophes" to ventilate what summarizes orchestrating a lack of knowledge about any real threat in the notion that we are living in a "risk society":

"The risk society is thus not a revolutionary society, but more than that a catastrophic society."[177]

How could one better phrase, that theorizing in the social sciences is not theorizing about something that exists, but a most imaginary image social science thinkers construct around the social, a free floating idea the social scientist discusses, that does not analyze the social reality, but that speculates about creating a possible image as how they want to see the reality and that phrases its observations to give this image a sense of plausibility, a prosaic theory, a theorizing poem. And what is the poem suggestion, how to see the world's social: The social is a dooming catastrophe, an apocalypse that disguises its apocalyptic nature its inhabitants are therefore not able to realize, except some exquisite sociological thinkers, proving with their apocalyptic scenarios a lack of what the world needs, needs what these sociological thinkers believe the world must be concerned with, with the concerns of most critical sociology.

---

[176] Ibid, p 68
[177] Beck, U., (1992) Risk Society, *Towards a New Modernity*, Sage Publication, English translation, London p 97/98

It is this image, serving a gloomy feeling of sociologists and all those who share these sociological concerns, that considers this feeling as an accordingly doomy opposition against whatever he argues about, Beck provides with his notion of a "risk society" in a book full of examples for really any use proving this doomy feeling anytime and anywhere, from childhood, education, the labor market, work, the state, science, simply the world as a whole, everything is a dooming apocalyptic disaster, which gives this book its title.

Beck's book is therefore a true treasury for academics who wish to find any quotation truly about any social sphere, if one seeks for such kind of doomy insights that present the whole world of the social as a dooming catastrophe, as a gigantic failure, failures compared to all the ideals social science believe the world should be like, ideals which though all remind us of their origin in the very society they are so concerned about.

In short, social scientists like Beck, leftist sociologists above all, are concerned about their social science missions critically scrutinizing if the world obeys their ideals and with his notion of a "risk society" Beck succeeded in providing the political debates mainly among those leftist intellectuals at that time in Germany and since then around the world, those who were seeking intellectual food for a new diffuse visions beyond a world they find simply bad, with thought, with a theory that provides a scientifically upgraded moaning for their desire to better care about the nature, about the whole world, not only our little national "we" (therefore the idea of "cosmopolitanism"), but about humans and *all the evils everywhere*, beyond the dirty reality of growth, poverty and war, nuclear power and so on and on, phenomena they do not want to analyze but to take them all as mere illustrations of the same and one and only vision, a lacking ideal of a world gathering mankind as a well ordered unity, unlike such dividing notions a the one of class society, united to respond to a bad world of apocalypses.

As usual: The fact, that this "theory" in that sense also provided those branches of the political elites, who, unlike those of all the Thatchers and alike in the world, were about to take over the political power to ruling the world under notions that presented their political aims as a challenge of and a service to mankind to the nature and to combat all the other risks and evils, was not only a well-received effect the risk society notion received among all those politicians seeing themselves as the do-gooders. It was the moral public debate in the 1980s in Europe that inspired Beck and provided all the examples for his very Christian diagnose of a bad world, a public debate he—in reverse—provided with a scientifically upgraded version of their moral world views. His book became a bestseller, drowning—not only—the class theory in enjoying the gloomy feeling to oppose this bad world and it would not have needed the empirical proof of the exploding nuclear power station to make this book a global morals court, shared knowledge, knowledge acknowledged across the world, a book as a truly never ending repository of apocalyptic conceptions of the world, therefore rightly one of the most quoted books, not at all only in sociology, but across the social sciences, appreciated among all the critical social sciences in the world, acknowledged and quoted namely by those social scientists in the imperial world, who enjoy to bother the world with the moral headaches of critical intellectuals as a contribution to the world's social science knowledge.

The ephemeral progress of ephemeral knowledge

"Change" is what according to social sciences across all disciplines is *the essential nature of the social;* consequently, it is the nature of social sciences to ever change their theories.

Obviously, since theories social sciences create interpret the social through meta theories, such theories only re-interpret the social once new meta theories occur and the new knowledge they create is no knowledge thanks to any change of the social reality but of their theoretical interpretative framework, as in our two examples such as "The Risk Society" or "Knowledge Based Economy", which have gained the status of such a meta theory. Since then, social science theorizing accommodates its theories to such new meta theories and re-phrases them within the new categories the meta theories provide.

As it can be easily traced in the case of the "knowledge based economy" theorem, theorizing in the social sciences replaced all categories related to knowledge and rephrased them in the sense of the new concepts of knowledge provided by the knowledge-based economy theory. Since then, knowledge is considered as a means for the market and categories such as knowing are exchanged with categories such as competencies, signaling the functionalism of knowing for a beforehand fixed use. It is not the case, that before knowledge was defined in this instrumental view, knowing was not defined instrumentally. There was no essential change in the subordination of knowledge under the needs of the economy mediated via the labor market, this was always the case; what has changed was the way to see knowledge and the change was to abolish older ideals about knowledge being a means for the developments of individuals and the like. This, interpreting knowledge and the acquisition of knowledge has been re-considered by the knowledge-based economy theory considering knowledge as the essential of the market economy and thus subordinating the conceptualization of knowledge to the objectives the market economy was the only change. Social science thinking only attributes their changed views on the same object of social thought as a change of this object. Substantially it does not make any difference in the instrumentalization of knowledge, whether the same instrumentalization is presented under the notion of developing human's talents or as a function in an economy presented as aiming at knowledge, it only changes the idealist title under which it is defined as essentially the same.

Whenever, social sciences change their meta theories, all those "turns" they detect as a need to re-phrase their theoretical frameworks, they rephrase their whole set of their theories across all disciplines under variations of the idealizations of the very unturned essentials of the substance of their social thought.

Nonetheless, the difference in phrasing the same essentials is however worth noticing: While the ideal presenting knowledge as a means to develop humans talents contains a reminder of a difference between the nation state, ever presented in its idealized version of the society, the community of individuals, the shift towards the idealized thinking in the concept of the knowledge-based economy theorem is to identify both. While the older ideal discusses the "society's" objectives to serve the objective of individuals, the notion of a society model aiming at knowledge presents humans serving the knowledge-based society/economy as serving themselves while serving the economy. It is only this, but it is this difference, translating the existential dependence of people's life from the nation state social constructs into what humans existentially are aiming at. The knowledge-based economy theorem presents the economy, the human is existentially dependent on to survive, forcing him by the threat of losing his economic basis, his income, and to adjust his knowledge to the "competences" to alien purposes the labor market might reward with a job, the knowledge-based economy notion identifies all those subordinations of knowledge and the knowledge holder under alien purposes as what knowledge is all about. Knowledge that serves the objectives of alien purposes one must obey is, in the knowledge based economy, presented as cognitively mastering the world, that is, both are the same: Serving the market economy is serving the lives of humans.

Though implicitly already answered within the discussion of the above two theories, one can conclude from both theories, which gained the status of meta theories "framing" social science thinking, knowledge that gained the status of a new acknowledged theory, what makes a theory such a new meta theory, a theory—as social science phrase it—representing a "turn" in social science theorizing, how such "turns" in theorizing are related to the claim that at least changing meta theories are a response to a changing social reality.

The two above mentioned examples allow to answer both questions and also give some hints about what this implies for answering the question how global social science thought progresses.

Both theories have in common that they contribute substantial theoretical elements about the view on the society as a whole: The knowledge-based economy about what the economy as a whole is about, the risk society about the global challenges such risk societies are facing and the plea for a global policy agenda dealing with those "risks".

Having the proof for such visions in their mind both theories have little theoretical scruples to violate even any rule of social science theorizing to arrive at what they are aiming at to prove. Ignoring that knowledge does not even obey the essentials of a market economy the knowledge based economy presents the subordination and domestication of knowledge and education under the needs of the global business, the business world in the imperial world has detected as their new means to set the standards for global economic competitiveness as an economy that aims at the promotion of learning and knowledge. The risk society constructs via an enumeration of apocalyptic scenarios, which apply the result of their theory to any phenomenon it reflects on and detects in any of the detected disasters a disaster, constructing a world that consists of nothing but such "risks" and concludes from the chaos this theory finds just everywhere, a call for a new political agenda for the world's nation states to create Ordnung, for a cosmopolitan thinker, global Ordnung; a call that coincides very well with the discourses of major parts of the political, economic and intellectual elites in the imperial world about some new global missions for an imperial policy agenda, ever discussed as a service for the globe.

Social science theories gaining the status of a meta theory if they manage to critically remind the ruling elites in the imperial world of their final idealized political missions or to re-new these missions with their ever critical reflections, idealized missions these elites, including the academic elites, the resource of the nation state society from which they recruit all their political, economic and academic elites, are convinced and committed to that they are the missions they are finally working for with all their agendas ruling the world's people. The tensions between the critical nature of social science theories guiding social science theorizing, institutionalized in their independency and the academia as the recourse of the society for the political elites is the institutionalized social conscience that insists against the social reality on the ideals this social reality desires to see as its missions.

Theories which manage to become the critical conscience of the imperial mission of imperial societies, ideally shared by their victims, reach the top of the social science hit lists. The "knowledge-based society" and "risk society" deserve their positions on this global hit list, as belonging to the most quoted theories providing the theoretical pre-assumptions through which social sciences across the world construct their theories, since they contribute some substantial elements to the most shared moral visions the elites of the imperial world appreciate as interpreting what they are doing.[178]

---

[178] Just to give briefly one other example how social science knowledge progresses by adjusting theories towards such theories framing a new wave of theorizing: "The rise of a cultural conception of knowledge is rooted in contemporary existence, in the current transition to a knowledge society. Today, at the start of the twenty-first century, it is argued by many that we are well on the way to an era beyond modernity and the sort of industrial economy and nation state societies that came with it; the terms suggested to refer to the transformations and the new type of system involved include post-industrial society, postmodernity, information society, risk society, globalisation and knowledge society. Though knowledge and information appear only in some of these terms, nearly all accounts suggest that issues of

To become a globally ruling theory one needs to create a theory that serves critically observing if imperial nation state rationales are rationales aiming at being a service for the world's mankind. It is this tension the mission of social sciences reminding the world of its ideals, which also implies conflictual relations between them and the political elites, though they only idealizes their rationales.

Consequently, another most critical theory complements another element, if not the core element, what the political, economic and intellectual elites in the imperial world, and certainly not only there, like to believe about the power they execute. The globally most quoted theories are Foucault's theories about political power. Compared to his theory, presenting political power, eagerly discussed along practices of tortures, as the desire for violence against the powerless, the aimless tautology of power for the sake of power, makes the ordinary execution of power via the discreet power of law of democratic nation states appear as the domestication of what political power is. Needless to say, that such theories about

---

knowledge and information are central to the transformation. Thus, whatever else the new era brings—the decline of the nation state, the globalisation of risks or individualisation—we are also entering a period focused upon knowledge and information (and these are entangled with the other processes). The *concepts* of epistemic culture and knowledge culture belong to this transformation."
K. Knorr Cetina, *Culture in global knowledge societies: knowledge cultures and epistemic cultures,* 'Knowledge cultures', in The Blackwell Companion to the Sociology of Culture, (ed. M. Jacobs and N. Weiss Hanrahan); 2005, Oxford, Blackwell.
Combining two of such framing theory bodies, the notion of the knowledge based society and the notion of culture, spiced with a pinch of "globalisation", saying that "all accounts suggest that issues of knowledge and information are central to the transformation", with the coquettish comment "whatever else the new era brings", the question if all those theories, categorically framing what the new theory about "epistemic culture" will create as a new theory about science, if all those theories framing theorizing are saying anything essentially new, not to mention if they are any true theories, the free speculative fabrication of a set of new social science theories about "epistemic culture" can start. The pre-fabrications for a new theory are prepared to find the data through the categories the framing theories provide, allowing to construct a theory about "epistemic culture" as a contribution to the new wave of theories framed by the notions of culture and a knowledge based society. A progress of knowledge about scientific thinking is made. To avoid a misunderstanding, this is not a critique of the theory, it is only an example illustrating how social science knowledge progresses. Discussing the theory is another matter.

political at the same time discredit the political power and are most appreciated among academic thinkers, who consider the missions as professional thinkers to control the political power if it fails to serve its mission to serve the inhabitants of nation states with their power over them.

To sum up: The progress of social science thinking is to critically trace if and how the changing policy agendas serve the idealized nation state rationales, idealized nation state agendas that found social science theorizing, thus ever critically ennobling them as finally aiming at these ideals, mostly critiquing them for failing to serve those objectives social sciences with these critical observations interpret into them as their final mission. Social science theories progress by forming legends fueled by the debates among the political elites about how to progress with the agendas of nation state and the economy it supervises, critically observed and interpreted by the social sciences as *the conscience of the very nation states ideals*.[179]

With this mission the need for this kind of social science knowledge progress never ends. That the demand for creating and ever updating ideals about the final aims of nation states never ends, this is the service the execution of the real nation state rationales, executed by the elites in the imperial world, provides to

---

[179] As three other examples for how social science knowledge progresses with the creation of new notions critiquing policies and thereby serving them, the notions of "social capital" and "social exclusion", then later "precariat" might serve. It was the left thinker Bourdieu, who in one swoop expropriated the wealthy people of the world with a new definition of what wealth is. Inventing the notion of "social capital", from one day to the next made the have-nots of the world to the real wealthy people, since they own, what all those people, only owning material wealth, never get, "social capital". Then, after the political elites, probably mainly in Europe, found that the notion of poverty must be a phenomenon of the previous century in a world that provides people who work throughout the whole life and remain being have-nots with social benefits can no longer be called poor, thanks to the categorical preparatory work of Bourdieu another, certainly also distinguished social scientist, provided the new category to define poor people as people, who were not lacking money, but social capital, and called this new problem of the have-nots "social exclusion". From then on poverty was no longer on in social science theorizing and replaced by a new ideal, later further developed to the notion of "precariat". Not only these notions entered and guided not only social science thinking, but from then on populated the public debates, preferably those among the political elites, harshly critiqued by the social sciences to not seriously care enough about the socially excluded precariat.

ever inspire to the criticism of social sciences, ever more or less disappointing the idealist thinkers with their real policy agendas. The idea, that is this nation state societies and all its constructs that cause all the "problems" for which solving them they elaborate new nation state visions, does not come to a social science mind, since this requires to think beyond their theories, their modes of thinking and especially to think beyond their idealized world visions, founding their thought.

Chapter E
Going beyond the social sciences

One can conclude: Social sciences do not create knowledge, insights about the social such as the natural sciences create about the nature; social sciences create knowledge which are rather a multiplicity of interpretations of the social, interpreted through modelled views, interpretative frameworks through which they construct their interpretations of the social—and this is, as the social science gurus confirm, not a failure, but what social sciences not only must do, but creating this multiplicity of possible interpretative views is for social sciences the nature of social thought. Creating knowledge, as they say, is a human hubris, a megalomania about science, and, as they finally managed to prove along their progress of knowledge, creating secured knowledge is not even what the natural sciences do—though not only the social science gurus practically trust the knowledge of the natural sciences as objective knowledge, when they drive with their cars to their universities having no clue how they work. Social sciences are a creature of the society that creates and sustains social sciences, they are a construct of a society that owns a science system with a particular social subject, monopolizing scientific thinking, academics, a state run science system with institutions, universities, sustained with a relevant part of its wealth, all creating knowledge for a society that though does—for good, or better, bad reasons—not know what the society is.

Nation state societies, societies constructed and sustained by nation states, economically reproducing the society via market economies, shape the social as the conflicting relations of competing humans, competing about what they contribute and gain from the wealth they produce, regulated and supervised by nation states with a power monopole defining via the rules of the competing subjects what they must do. Excluding each other from what they need to benefit from their own properties, the nation state humans, bound to exist from what they own as a means for their live aims, are for the sake of their own reproduction within this society bound to look at the social world as a means while striving for their aims as privates: From this practical view of competing private property owners the world is seen as how it serves to benefit from what they own. It is this practical perspective seeing the social world, interpreting the social world as means for their private purposes that scientific thinking serves, presenting the social and the natural world as utilities that are made for the final purpose to serve the live rationales of the nation state human's subjects, the competing private property owners.

Accommodating theorizing towards this utilitarian view that must interpret the social as a means for the private purposes is the view social sciences transform into a mode of scientific thinking, that results in interpretations of the social, interpreted through the ideals founding and framing social science thinking. It is the politically set free materialism of the free and equal subjects of this society, offering its inhabitants to utilize the nation state social constructs and the nature as a means to achieve their interests as privates that social sciences insert into their views to seeing the objectives of nation state societies constructs as serving its inhabitants, that constitutes social science and their approach to thinking as the institutionalized mission they have for the society.

This thinking does not create any secure insights into the society but a multiplicity of optional interpretations of what the society may be. However, this *is* though the kind of knowledge this society established and sustains, as one can already conclude from the fact that no relevant member of the society complains that a science, which does not create knowledge about the society, is a failure and a waste of resources and instead insists on social sciences to create verifiable knowledge knowing what the society is. This society, sometimes called knowledge based society, does obviously not need knowledge about the social. Knowledge about the social, knowledge that knows how this society, its politics and economy are working, is of no use for how this society works. Holding social sciences which create different interpretations of the social seemingly satisfies the knowledge this society wants. It is a society that sustains social sciences, which establishes a way of social thought, that unlike its predecessors, do not create speculative ideas, presenting the reality as the realization of ideas, it is science which creates theories that are scientifically elaborated reconstructions of the real world, reconstructed through images imagining how the real world is supposed to be. This, scientifically elaborated reconstructions of the real world, reconstructed through images imagining how the real world is supposed to be is the social science knowledge this society obviously desires and therefore institutionalizes.

It is these social sciences which create this kind of relative knowledge that serves as the knowledge needs of the society members, more precisely it serves those who acquire this social science knowledge. Knowledge that presents the social as a variety of possible interpretations of the social reality cannot guide anybody's social practices. This knowledge serves as an offer for a variety of ex post interpretations of commanding actions, which obey other instructions than the knowledge the social sciences articulate for their ex post interpretations.

It is knowledge reserved for those who have the commands in this society. Ordinary citizens never get in touch with social science knowledge, the command receivers receive "competences", practical knowledge executing alien purposes which are not their business. The social science knowledge presenting the commands of the elites as a service for the commanded is reserved for the commanders. It is the knowledge for the political, economic and academic elites, who also do not need to know how this society, its political system and its economic reproduction work for their commanding actions. It is rather the other way round, knowledge that creates idealized images, speculative interpretations about the society, the economy and the politics they command, images which interpret their commanding actions as serving the people they command, is the knowledge they own and appreciate. The knowledge that guides what they practically do is knowledge they do not get from the social sciences. Except for interpreting the social as a useful means for the society members, the knowledge social sciences create is also useless for the practical work of the commanders, however, this, knowledge for the ex-post interpretation of what the whole society is ideally striving for, is what it is useful for and for this very use it is institutionalized as the sciences system of this society.

This social science model of scientific thinking overcomes the subordination of thinking under given ideas as this was the case in the classic philosophies, which present the world as the realization of a presupposed philosophical telos. The social science model of scientific thinking rightly critiques the speculative knowledge of the classical philosophies derived from a telos, and, instead presupposes the real world as a product of humans and makes the social reality the object of thinking.

> "Our researchers, then, in every branch of knowledge, if they are to be positive, must be confined to the study of the real facts without seeking to know their first causes or final purposes."[180]

"To be positive" it replaces knowledge, classical philosophies derived from speculative ideas, by scientific thought subordinated under the social objectives incorporated in the objects of thinking. Its way of cognitions, making the objects of thinking, "the real facts" the proof for the pre-assumptive thought through which they theorize, enthrones the ideals of philosophies by an idealized reality as ruling thinking.

It is this mode of thinking in the social sciences that creates their idealized knowledge, that is not founding what and why the inhabitants of this society do what they do, but knowledge which creates—in order "to be positive"—presupposing images materialized in their meta-theories about the society and its social constructs through which they theorize, presupposing that they are finally aiming at serving its citizens, whose live is not crafted by the knowledge the social sciences create, but by the real power of the social constructs ruling their lives beyond any insights, the least by those idealized views the social sciences have on the social. It is a science that creates knowledge that exists beyond the rationales of the life of the privates and carrying out the rationales of the life of privates does not need to know any knowledge these social science create. They create knowledge—mainly for the social sciences, the recruitment pool for the political and economic elites, assuring themselves that their execution of political and economic power finally aims at the ideals the social sciences ever critically confirm.

---

[180] A. Comte, (1998) *Introduction to Positive Philosophy,* edited by Frederick Ferre, Indianapolis p 21. Interpreting Comte's' notion about "positive philosophy" as if Comte was an early positivist, is a pretentious misunderstanding, mixing up the theorizing technologies, namely in sociological thinking, the variations in the methodological apparatus to operationalize social sciences in the established social sciences and their discourse about how to practice their scientific scepticism as science, with Comte's attempts to describe a distinctive way of theorizing between theorizing in the classical philosophies and theorizing in the emerging social sciences of the emerging nation state society.

It is this concept of social thought the imperial societies invented and exported across the world and that, thanks to the fact that the opposition in the former colonized world did not know any other opposition against the imperial world but to imitate the society and economy model of the imperial world, and with the imitation of the society system including its concept of social thought the social sciences became the global practice of social thought, not at all only in the imperial part of the world. This is how the social sciences think about the world's social—and how they critically think since 200 years about war and poverty in a world that is still a world of war and poverty.

Postscript

Forced by the insurrection of the feudal people the ruling elites responded by acknowledging the ruled peoples free will *and* by dictating how and towards which aims to execute this will, they created a nation state of citizens; the nation state made knowing the exclusive matter of the free will of a selected species of thinkers, thinkers, who use this freedom to critically celebrate the nation state as aiming at serving the free will of its subordinated citizens and to discuss the risks of their free will as a challenge for the nation state to domesticate their will. This is what social science thinking is about. There might be other ways for social thought beyond social science thinking.

Thereby everything what needs to be done to get out of social science thinking and has been basically said.

It is obvious that thinking beyond social sciences is thinking that

- theorizes through no presuppositions, that eliminates any "context" from theorizing;
- it questions the explanatory power of the disciplinary meta theories and categories before using them as knowledge guiding the creation of theories;
- It is thinking that, as a consequence, thinks beyond the disciplinary architecture and its divisions of social thought and its explanatory models. It is neither inter—, trans—nor any way of disciplinary thinking, it is thinking that is not thinking in any pre-constructed thinking compartments;
- It is thinking that does not need methodologies, imposing into the objects of thinking what they intend to find, and it does not discuss how it arrived at what it says, but it discusses what it says.
- Discussing the theories this knowledge creates, does not argue about any of means and circumstances under which it has been created; it discusses the coherence of its social thought, where ever about what ever;
- Social thought beyond social science thinking does not know any spatial attributes or borders nor any definitions

of what scientific knowledge is. It is social thought to be created by anybody and anywhere and it is discussed if it is knowledge about the social or not.

Thinking and arguing beyond the social science dogmas about science and having scientific discourses is theoretically as uncomplicated as thinking originally is; it is though not that uncomplicated, because it requires nothing less but a sort of scientific revolution, liberating social thought from the dogma of the social sciences, that social thought must be preoccupied thought, a dogma that disempowers humans knowledge and bends there thought under the rationality of political power. Going beyond social sciences is in the first place nothing but critiquing the existing social science theories, tracing their false thought originating from their pre-assumptive thinking. And this is the one and only way to erode and from there to go beyond social thought about the world's social under the regime of the social sciences.

***ibidem*-Verlag**

Melchiorstr. 15

D-70439 Stuttgart

info@ibidem-verlag.de

www.ibidem-verlag.de
www.ibidem.eu
www.edition-noema.de
www.autorenbetreuung.de